Grassland Management

Grassland Management

Edited by

Stephanya Lynn JonasLabee

Delve PUBLISHING

www.delvepublishing.com

Grassland Management

Stephanya Lynn JonasLabee

Delve Publishing
224 Shoreacres Road
Burlington, ON L7L 2H2
Canada
www.delvepublishing.com
Email: orders@arclereducation.com

ISBN: 978-1-77407-713-9 (Hardcover)

ABOUT THE EDITOR

Stephanya Jonas-Labee has been passionate about the environment her whole life. After completing her bachelor's at UCSC, she moved to Amsterdam to earn her Master's of Earth Science with a focus on environmental management and sustainability from the University of Amsterdam. When not reading about new sustainable technologies, she enjoys traveling the world, trying out new vegan recipes, and working on her garden.

TABLE Of CONTENTS

GLOSSARY

A

Abandonment - The action or fact of abandoning or being abandoned

Abrupt - Sudden and unexpected

Agrarian - relating to cultivated land or the cultivation of land

Antelope - A short-horned goat antelope native to mountainous terrain in southern Europe

Anthropogenic - originating in human activity

Aquifer - An aquifer is an underground layer of water-bearing permeable rock, rock fractures or unconsolidated materials

B

Bioclimatic - relating to the interrelation of climate and the activities and distribution of living organisms

Biodiversity - he variety of plant and animal life in the world or in a particular habitat, a high level of which is usually considered to be important and desirable.

Bioindicators - an organism used as an indicator of the quality of an ecosystem, especially in terms of pollution.

Biomass - the total quantity or weight of organisms in a given area or volume.

Bushing - a shrub or clump of shrubs with stems of moderate length

C

Compaction - the exertion of force on something so that it becomes more dense

Continent - is one of several very large landmasses. Generally identified by convention rather than any strict criteria, up to seven regions are commonly regarded as continents

Crevice - A narrow crack or fissure, as in a rock or wall

Cultivars - A plant variety that has been produced in cultivation by selective breeding. Cultivars are usually designated in the style Taxus baccata 'Variegata'

D

Decadent - characterized by or reflecting a state of moral or cultural decline

Decaying - rotting or decomposing through the action of bacteria and fungi

Deferment - An act or instance of deferring or putting off

Defoliated - remove leaves from (a tree, plant, or area of land), for agricultural purposes or as a military tactic

Desertification - is a type of land degradation in drylands involving loss of biological productivity caused by natural processes or induced by human activities

Distinguishing - characteristic of one thing or person, so serving to identify it; distinctive

E

Ecosystem - a biological community of interacting organisms and their physical environment

Edaphic - is one of two main divisions of soil science, the other being pedology

Emancipation - The act of setting free from the power of another, as from slavery, subjection, dependence, or controlling influence

Erosion - the process of eroding or being eroded by wind, water, or other natural agents

Evapotranspiration - is the sum of evaporation and plant transpiration from the Earth's land and ocean surface to the atmosphere

F

Feedlots - is a type of animal feeding operation which is used for the efficient raising and finishing of livestock, notably beef cattle, but also swine, horses, sheep, turkeys, chickens or ducks, prior to slaughter

Fertilization - also known as generative fertilization, insemination, pollination, fecundation, syngamy and impregnation, is the fusion of gametes to initiate the development of a new individual organism or offspring

Fescue - A long, thin piece of metal on a scale or dial which moves to indicate a figure or position

Fluxes - is a chemical cleaning agent, flowing agent, or purifying agent

Foothills - a low hill at the base of a mountain or mountain range

Forage - obtain (food or provisions) by searching

G

Grassland - areas where the vegetation is dominated by grasses; however, sedge and rush families can also be found along with variable proportions of legumes, like clover, and other herbs

Grazing - is a method of animal husbandry whereby domestic livestock are used to convert grass and other forage into meat, milk, wool and other products, often on land unsuitable for arable farming

H

Habitat - the natural home or environment of an animal, plant, or other organism

Harmonization - The fact or process of producing a pleasing visual combination

Harrowing - acutely distressing

Heralding - Present participle for to give a slight indication of beforehand

Herbaceous - denoting or relating to herbs (in the botanical sense)

Herbivorous - feeding on plants

Herders - is a worker who lives a possibly semi-nomadic life, caring for various domestic animals, in places where these animals wander pasture lands

Herding - is the act of bringing individual animals together into a group, maintaining the group, and moving the group from place to place—or any combination of those

Heterogeneity - the quality or state of being diverse in character or content

Hoeing -use a hoe to dig (earth) or thin out or dig up (plants)

Homogeneous - denoting a process involving substances in the same phase

Hydrology - the branch of science concerned with the properties of the earth's water, and especially its movement in relation to land

Hypomagnesaemia - is a condition in which someone has abnormally low levels of magnesium in their blood

I

Intercepted - obstruct (someone or something) so as to prevent them from continuing to a destination

L

Legumes - a leguminous plant (member of the pea family), especially one grown as a crop

Livelihoods - refers to their "means of securing the basic necessities of life"

Lumpy - Not having a level or smooth surface

Luxuriant - Rich and profuse in growth

M

Manifestation - The action or fact of showing something

Mediterranean Sea - is a sea connected to the Atlantic Ocean, surrounded by the Mediterranean Basin and almost completely enclosed by land

Mongolian - the language of Mongolia, a member of the Altaic family with an unusual vertical cursive script

Moths - an insect with two pairs of broad wings covered in microscopic scales, typically drably coloured and held flat when at rest.

Mowers - a machine used for cutting grass.

N

Nibble - To eat reluctantly and in small bites.

Noxious - Harmful, poisonous, or very unpleasant

O

Overgrazing - excessive grazing which causes damage to grassland

P

Palatability - is the hedonic reward provided by foods or fluids that are agreeable to the "palate", which often varies relative to the homeostatic satisfaction of nutritional, water, or energy needs

Pampas - A large area of flat land with few trees

Pastoralists - is a form of animal husbandry, historically by nomadic people who moved with their herds

Pasture - pasture lands in the narrow sense are enclosed tracts of farmland, grazed by domesticated livestock, such as horses, cattle, sheep, or swine

Pastures - land covered with grass and other low plants suitable for grazing animals, especially cattle or sheep

Peculiarities - a strange or unusual feature or habit

Penetrate - go into or through (something), especially with force or effort

Perturbations - anxiety; mental uneasiness

Phenology - is the study of periodic plant and animal life cycle events and how these are influenced by seasonal and interannual variations in climate, as well as habitat factors

Ploughing - turn up the earth of (an area of land) with a plough, especially before sowing

Prairie - are ecosystems considered part of the temperate grasslands, savannas, and shrublands biome by ecologists, based on similar temperate climates, moderate rainfall, and a composition of grasses, herbs, and shrubs, rather than trees, as the dominant vegetation type

Precipitation - is any product of the condensation of atmospheric water vapour that falls under gravity from clouds

Puszta - Plains; A vast wilderness of shrubs and grassland

R

Rangelands - are grasslands, shrublands, woodlands, wetlands, and deserts that are grazed by domestic livestock or wild animals

Rehabilitation - the action of restoring someone to health or normal life through training and therapy after imprisonment, addiction, or illness

Replenished - restore (a stock or supply) to a former level or condition

Resilient - able to withstand or recover quickly from difficult conditions

Resuscitate - Revive (someone) from unconsciousness or apparent death

Ruminants - an even-toed ungulate mammal that chews the cud regurgitated from its rumen. The ruminants comprise the cattle, sheep, antelopes, deer, giraffes, and their relatives

S

Savanna - is a mixed woodland-grassland ecosystem characterised by the trees being sufficiently widely spaced so that the canopy does not close

Serfs - A partially free peasant of a low hereditary class

Silage - grass or other green fodder compacted and stored in airtight conditions, typically in a silo, without first being dried, and used as animal feed in the winter

Steppes - a large area of flat unforested grassland in south-eastern Europe or Siberia.

Stockpiling - accumulate a large stock of (goods or materials).

Stratification - refers to society's categorization of its people into groups based on socioeconomic factors like wealth, income, race, education, gender, occupation, and social status, or derived power

Succumb - Fail to resist pressure, temptation, or some other negative force

Surge - A sudden powerful forward or upward movement, especially by a crowd or by a natural force such as the tide

Susceptible - likely or liable to be influenced or harmed by a particular thing

Synergistic - relating to the interaction or cooperation of two or more organizations, substances, or other agents to produce a combined effect greater than the sum of their separate effects.

T

Tannins - a yellowish or brownish bitter-tasting organic substance present in some galls, barks, and other plant tissues, consisting of derivatives of gallic acid

Temperate - the temperate or tepid climates of Earth occur in the middle latitudes, which span between the tropics and the polar regions of Earth

Thrive - grow or develop well or vigorously

Tillage - the preparation of land for growing crops.

Trampled - tread on and crush

V

Venation - the arrangement of veins in a leaf or in an insect's wing

Vigor - physical strength and good health

Vole - A small, typically burrowing, mouse-like rodent with a rounded muzzle, found in both Eurasia and North America.

W

Widespread - found or distributed over a large area or number of people

Z

Zebus - sometimes known as indicine cattle or humped cattle, is a species or subspecies of domestic cattle originating in South Asia

LIST OF FIGURES

LIST OF TABLES

LIST OF ABBREVIATIONS

ATP	Adenosine Triphosphate
CAP	Common Agricultural Policy
CCC	Canadian Climate Centre
CFS	Canopy Shade Factor
CHO	Carbohydrate
CO_2	Carbon Dioxide
CSC	Conservation Significance Category
DM	Dry Matter
FAO	Food and Agriculture Organization
FLPMA	Federal Land Policy Management Act
FVC	Green Vegetation Cover
FVS	Floristic Values Score
GCM	Global Climate Models
GFHI	Geophysical Dynamics Laboratory High Scenario
GHCS	Grassland Household Contract System
HI	Harvested Index
ICAR	Indian Grassland and Fodder Research Institute
LAI	Leaf Area Index
LCCS	Land Cover Classification System
MARS	Monitoring Agriculture by Remote Sensing
MIG	Management Intensive Grazing
NADPH	Nicotinamide Adenine Dinucleotide Phosphate
NDVI	Normalized Difference Vegetation Index
NIRS	Near Infrared Reflectance Spectrometry
NPP	Net Primary Production
OM	Organic Matter
PAR	Photosynthetically Active Radiation

PPT/PET	Potential Evapotranspiration Rate
RUE	Radiation Use Efficiency
SCOPE	Scientific Committee on Problems of the Environment
SOC	Soil Organic Carbon
SOM	Soil Organic Matter
USSR	Union of Soviet Socialist Republics

PREFACE

A major part of the landmass of the earth is covered with grasslands. This adds to the scenic beauty of a place as well as provides a rich diversity in the ecosystem. The grasslands are of great importance to the people for their living as well as for animals to graze upon and to find their habitats.

The grasslands need to be managed properly so that they can serve the purposes they are meant for. The recent pressure on the grasslands by the various human activities and some natural disasters has rendered them barren in some parts of the world. This has adversely affected both human and animal lives in those regions.

This book aims to throw light on the various aspects of grassland management by indicating the role of the various organisms in it. Grassland management is gaining importance gradually around the world however there is limited literature on this subject as of now. One main purpose of this book is to develop awareness among the masses about the need to manage the grasslands sustainably and to provide them with knowledge and information that they can refer to.

The book brings awareness about the meaning of grasslands and its various types and characteristics as well as the position they hold in the environment. It also discusses the various kinds of agricultural practices and nesting practices that are going on in the various grasslands.

The book highlights the production of grasslands, that is, the way in which the plants and grass grow and the practices involved in growing them efficiently. It also throws light on the various kinds of conservation practices for the grasslands.

There is also discussion about impact that climate change is having on grasslands in various parts of the world. The book addresses the various techniques and methods that are employed in the effective management of grasslands, also mentioning the principles of grazing management and emphasizing the role that grazing plays in grassland management.

Moving further, the book talks about the various policies and laws that have been formed to manage the grasslands in different parts of the world, bringing to focus the manner in which the grasslands are classified around the world.

The book also brings to focus the improvement and rehabilitation of the grasslands and degraded areas and the various social, economic and environmental factors that contribute to it. This is followed by the information on the effect of grassland management on the production of different kinds of plants and the various environmental benefits that can be obtained by effective management of grasslands.

I hope that this book is sufficiently informative on the subject of grassland management and that reading this book can provide the necessary motivation and knowledge to conserve the grasslands.

1

Introduction to Grassland

CONTENTS

Grassland is an area that contains or is covered with grass, including wild grasses and the trees. The grassland is mostly covered by vegetation cover. However, human activities on the land have adversely affected the quality of the land, thereby impacting the health of the ecosystem.

This chapter starts with the introduction to grasslands, the different types, and the characteristics of grasslands.

The chapter goes on to explain the types of grassland habitats in the new hemisphere and the changes in the grassland habitats from earlier times. Finally, there is an emphasis on climate vulnerabilities of the grasslands, nesting birds of the grasslands, and grassland management.

1.1. INTRODUCTION

Grassland is considered as an area in which the vegetation is dominated by around continuous cover of grasses. Grasslands tend to occur in environments that is conducive to the growth of this plant cover but not to that of the taller plants, specifically trees and shrubs. There are various factors that prevent the establishment of such taller, woody vegetation.

Grasslands are said to be the one which is the most widespread of all the major vegetation types of the world.

However, this is due to the human manipulation of the land which has significantly altered the natural vegetation, further creating the artificial grasslands of the cereal crops, pastured and some other areas which do require some form of repetitious as well as unnatural disturbances like the cultivation, heavy grazing, burning or mowing to persist.

There are many names given to the grasslands, such as prairies in North America, savannahs and veldts in Africa, Asian steppes, Australian rangelands, and pampas, llanos and Cerrado in South America.

But they are all places where there is not enough rain to support the growth of trees.

These lands are instead covered in grasses as well as grass - like plants that have growing points close to the soil and can keep on growing despite being nibbled on by animals.

These kinds of grasses can provide support to the high densities of grazing animals such as antelope, bison, and zebra. These herds provide support to the iconic predators which include lions and cheetahs.

Figure 1.1. Grassland Grass Meadow Silver.

Source: Image by Pixabay

Grassland ecosystems are specifically fragile due to the scarcity of water. In Australia, Africa, and South America, grasslands are dependent on regular episodes of fire for renewal. These areas are prime targets for the purpose of human development and so, are often exploited.

Humans tend to plow grasslands to plant wheat and some other crops, replace wildlife with the domestic livestock, and kill predators and prey alike. There are a few grasslands that are protected from development.

When the European settlers came to North America, much of the grasslands of the continent were quickly transformed into agricultural fields, which in turn wiped out some of the native wildlife species such as the bison.

Then the consequences further obtained were devastating. In the early 1930s, drought struck the region, crops could not grow, and bare fields were swept into powerful dust storms. This destroyed the farms as well as made the land uninhabitable for a number of years.

1.2. ORIGIN

Natural grasslands can be thought of as intermediates in an environmental gradient, keeping forests at one end and deserts at the other. The most favorable environments are occupied by the forests, where the moisture is adequate for the purpose of growth and survival of tall, dense vegetarian which is dominated by trees. Deserts are mainly found in the places where the moisture is lacking and continuous, permanent vegetation cover cannot be maintained. Grasslands tend to lie in between these two extremes.

Similar to savannas, grasslands arose in the deserts and scrublands (into which they commonly blend) during the period of cooling and drying of the global climate, which occurred during the Cenozoic Era (which is 65.5 million years ago to the present). Indeed, the grass family has evolved in the early years in this era (Poaceae or Gramineae). The date of the earliest appearance of grasslands tends to vary depending on the region.

Considering a number of different regions, there is a succession of vegetation types which can be recognized in the Cenozoic fossil record, along with the drying out of the climate progressively.

For example, during the past 50 million years in central Australia, tropical rainforest gave way successively to savanna, grassland, and ultimately desert. In some areas, the expansion of the grasslands took place while approaching their modern extent. This occurred only during the extremely cold, dry intervals which are referred to as ice ages in the north temperate regions of the past two million years.

There is a dynamic balance that commonly exists between grasslands and related vegetarian types. At some times, droughts, fires, or episodes of heavy grazing favor grassland and at other time, there are wet seasons and an absence of considerable disturbances which favor woody vegetation. There can be changes in the severity or frequency of these factors which can cause a change from one vegetation type to another.

There is another grasslands type that occur in places that are too cold for trees to grow, beyond the forest limits of high mountains, or at high latitudes. This characteristic type of grassland, tussock grassland, is found in cool and moist parts of the Southern Hemisphere at various latitudes. This region is dominated by tussock or bunch grasses which in turn develop the pedestals of matted stems, giving the vegetation a lumpy appearance. They are found in the tropics above the forest limit on some high mountains for example, in New Guinea and East Africa. Considering the higher latitudes of the Southern Ocean, they tend to form the main vegetation of sub-antarctic islands. Typically, they are also located in the drier, colder parts of New Zealand and the southernmost regions of South America.

However, not all-natural grasslands arise from the circumstances related to climate. In certain areas, there can be woody plants that have limited growth for some reason, further allowing the grasses to dominate. There are several causes, one of them is seasonal flooding or waterlogging, which is responsible for the creation and maintenance of large grasslands in the parts of highly seasonal subtropics and also, in smaller areas of some other regions.

For example, one of these seasonally flooded subtropical grassland is the Pantanal in the Mato Grosso region of Brazil. Dry grasslands prevail for half of each year and shallow wetlands, across an area of 1,40,000 square kilometers (54,000 square miles), along with the small forest patches which are restricted to low rises which do not flood during the wet season. There are many other areas where the climate is suitable for forest growth, very shallow or infertile soils can prevent tree growth as well as lead to the development of grassland.

The largest areas of natural grassland, which result from climate dryness, can be classified into two broad categories. One type is tropical grasslands, which generally lie in between the belts of tropical forest and desert. Another type is temperate grasslands which generally lie in between deserts and temperate forests.

Tropical grasslands tend to occur in the same regions as savannas and the difference between these two vegetation types is rather arbitrary. This depends on whether there are a few or many trees. Similarly, temperate grasslands can have a scattering of shrubs or trees which in turn blurs their boundaries when they occur adjacent to scrublands or temperate forests.

Tropical grasslands are majorly found in the Sahel south of the Sahara, in East Africa as well as in Australia. Principally, temperate grasslands occur in North America, Argentina and also, across a broad band from Ukraine to China. But there are a number of regions which have been substantially altered by agricultural activities.

There are a number of grasslands that were formerly assumed to be natural. They are now recognized as having once been forests that tend to grow in a marginally dry climate. Human disturbance in the early times is responsible for their transformation. For example, almost the whole extensive lowland grasslands found in the eastern part of the South Island, New Zealand are said to have been created by forest - burning which was carried out by the Polynesians. Polynesians are considered to be the country's first colonists during the eight centuries before the European settlement in the 18th century.

Seminatural grasslands can occur where woody vegetation was once cleared for the purposes related to agriculture which have been abandoned.

There is a return to the original vegetation which is prevented by repeated burning or grazing. These types of grasslands can be very dense in wet tropical regions. There can be one example of those which are in East Africa and further dominated by elephant grass (*Pennisetum purpureum*) or in New Guinea by pit-pit grass (*Miscanthus floridulus*). Both of these grow 3 meters (9.8 feet) tall.

1.3. BIOTA

Grasslands are spread across a large portion of the continent pre-European North America, ranging from the Rocky Mountains in the west to the deciduous forests in the east. Considering this vast expansion of the grasslands, there are only tiny fragments which remain in certain condition remotely. This is similar to their original state. There is the largest central area which contains mixed prairie, further dominated by a number of different species of grasses namely *Stipa*, *Agropyron*, *Bouteloua*, and *Koeleria*.

In the North, mixed prairie gave way to a fescue prairie along with *Helictotrichon* and *Festuca*. In the west, it gave way to a short-grass steppe which is dominated by *Buchloe dactyloides* and *Bouteloua gracilis*. In the east, it gave way to a tall-grass prairie along with the bluestem grasses *Andropogon gerardii* and *A. scoparium*. Generally, the trees and shrubs were absent, but a large variety of herbaceous plants occurred with the grasses.

Figure 1.2. Bluestem Prairie.

Source: Image by Flickr

There are a large number of grazing mammals of the North American prairies such as the bison, pronghorn antelope, and grey wolf. The badger and several rabbit and hare species were widely spread, among many small burrowing rodents. Grasshoppers are considered among the invertebrate fauna which were and still are particularly important. In the grasslands, typically, the total biomass of invertebrates exceeds that of the much larger and more conspicuous vertebrates, except in the regions which contain a large number of domestic stocks.

In South America, the principal region of grassland is in the southeast portion of the continent. It can be divided into the Pampas of Argentina as well as the compos of the adjacent areas of Uruguay and Brazil. *Stipa* is said to be the most diverse among the many kinds of grass in the Pampas, whereas there is another suite of grasses in the Campos which is included among its more common members *Andropogon* and *Paspalum*. The vegetation of the region is greatly altered by centuries of heavy grazing as well as burning. Before this, the principal large herbivore was the pampas deer, from which only one small herd survives. Armadillos, opossums, and rodents also considered to be abundant and were preyed upon by various predators such as jaguars, pumas, and foxes. The rhea is a large and flightless bird, which is indigenous to this area.

The Eurasian steppe tends to occupy an extensive tract of the former Soviet Union and Mongolia, and that is in a number of ways which are very similar to the prairie of North America and hence harbors many similar plants as well as animals. There are a number of different species of *Stipa* which tend to dominate the flora in most of the areas. They are mixed in spots with some other grasses, primarily *Festuca* and *Agropyron* .

There are some activities such as feeding and burrowing of rodents which are important for the maintenance and composition of the vegetation. These include the large marmots and a variety of voles and other smaller rodents. *Lasiopodomys brandtii* is a vole in Mongolia which in some years can consume such a high proportion of the vegetation that it converty its grassland habitat to a virtual desert.

The Sahel is considered to be the broad band of grassland which crosses the western and north-central Africa, south of the Sahara. The Sahara is said to be the largest area of tropical grassland. The millennia human populations have put a number of demands on the region, and therefore, its present condition is not so similar to its natural condition. The most common grasses in this area include *Aristida*, *Cenchrus*, and *Schoenefeldia*.

There are some other species that are highly palatable to grazing animals. Such kinds of species are now restricted to rocky sites that offer some protection. These species can have once been far more widespread as well as important. In places were shrubs and small trees occur, the vegetation would be more accurately referred to as a scrubland or savanna. These are not so easily transformed into grassland by practices like grazing, burning, and fuel gathering.

The wetter environments that exist in the Sahel are included in the grassland of East Africa and hence, are more diverse. In places where the forests have been destroyed, a tall grassland that contains *Pennisetum* or *Hyparrhenia* is developed. These grasslands can be kept in this condition indefinitely by the practices of burning or the grazing of some herbivores like elephants.

There are some other grasses like *Aristida* and *Chrysopogon* which are considered to be important in the drier sites and *Themeda* occurs in the cooler places at higher altitudes. There are herbivorous mammals such as wildebeest, different antelope species, buffalo, elephants, and rhinoceroses. Some of the carnivores include hyenas, mongooses, jackals, cheetahs, and lions.

In Australia, in the extensive arid areas, the tropical grasslands are generally dominated by species of the spinifex grasses, *Plectrachne* and *Triodia*. These form characteristic hummocks by trapping windblown sand at the bases of their tussocks. Grasslands in the moister, northern area are dominated by *Heteropogon* and *Sorghum*. *Astrebla* (Mitchell grass) is considered to be prevalent in the areas which are seasonally arid areas, especially on cracking clay soils in the east. There are some other grass species that are usually subordinate but can dominate in spots.

There are certain woody plants, specifically *Eucalyptus* in moister places and *Acacia* in arid areas. These plants can be so large in number that the vegetation cannot be considered true grassland. Historically, the Mitchell grasslands were much purer until they were altered by the practices such as heavy grazing of domestic stock. Today, there are vast tracts that have been invaded by the African shrub *Acacia nilotica*, which was introduced by humans.

In Australian grasslands, the largest native animals are kangaroos. Among these, the biggest one is the red kangaroo, which is characteristic of the dry inland areas. In these areas, natural grasslands are found. However, a number of mammals have been introduced from other continents and have become as common. These introduced species include domesticated stock, such as cattle and sheep, and a range of feral fauna such as camels, donkeys, horses, and goats.

European rabbits have also been introduced which are widespread, abundant, and highly destructive. There is one main predator of the Australian grasslands, the dingo or wild dog. There is a diverse range of reptile species in the area as well. Bird species include the emu, which are large and flightless, as well as a wide range of parrots.

The tussock grasslands of New Zealand, as well as the sub-antarctic islands, are said to be commonly dominated by species of *Poa*. There is related vegetation which also occurs on high mountains in equatorial regions. In most of the cases, these grasses have evolved in the absence of grazing mammals. There has been severe degradation of the vegetation in many places which are caused because of the introduction of exotic species such as deer and sheep.

1.4. ENVIRONMENT

There are grassland climates which do vary. But, generally, all the large regions of natural grassland are hot, especially in summer, and dry, but not to such an extent that deserts are. Generally, tropical grasslands receive 500 to 1,500 millimeters (20 to 60 inches) of rain in an average year. They experience temperatures of approximately 15 to 35 °C (59 to 95 °F) year round. The dry season may last as long as eight months.

During the wet season there is an excess of rainfall which in turn results in the ephemeral river flow. The tropical grassland climate overlaps very broadly with savanna. These vegetation types differ a little bit from each other and so, a savanna is said to be merely a grassland with scattered trees. A number of small changes occurring in management and usage can convert one to the other.

Temperate grasslands are considered to be much drier and colder in comparison to the tropical grasslands, at least for part of the year. There can be variations in the seasonal temperature which may be slight in tropical grasslands. This temperature may vary by as much as 40 °C (72 °F) in temperate grassland areas. In the North American grassland areas, the mean annual rainfall is 300 to 600 millimeters (__-__ inches).

In January, the mean temperatures range from −18 °C (0 °F) in the north to 10 °C (50 °F) in the south. In July, the corresponding values are 18 °C (64 °F) and 28 °C (82 °F). Considering most of the northerly areas of the North American grassland zone, the mean annual temperature is below 0 °C (32 °F).

Grasslands are related to a number of different types of soil across a wide range of climatic as well as geologic conditions. The soil formation is further influenced by the grassland ecosystem, and this differentiates the grassland soils from other soils.

Usually, the nature of grass litter and its pattern of decomposition leads to the development of a dark and organically rich upper soil layer. This layer can reach 300 millimeters below the surface. This layer is not ‘ present in desert soils and is different from the surface layer of rotting leaf litter typical of forest soils. In structure, it is friable and rich in plant nutrients. Typically, lower soil layers are pale and yellowish in color, especially at depths around two meters.

1.5. POPULATION AND COMMUNITY DEVELOPMENT AND STRUCTURE

Natural grasslands can be either tropical or temperate and tend to occur in environments that have a short season of favorable growing conditions. This growing season in tropical regions is usually the rainy season. But in some cases, the growing season is when the ground is not waterlogged or even submerged. In temperate grasslands the growing season is usually the short period in between the cold, damp winter and the hot, dry summer.

Perennial grasses tend to rely on subterranean reserves of stored nutrients for the purpose of rapid shoot growth. These grasses are well adapted in order to exploit such a brief growing season, quickly reaching their maximum size and completing their seeding within a few weeks. Their aboveground parts then die, providing the potential fuel for the grass fires which typify these environments.

However, the underground perennating roots and rhizomes of the grasses are relatively well-protected from fire. The grasslands produce large amounts of new growth if they are subjected to some type of repeated disturbance, such as grazing or fire, and this tends to prevent the accumulation of a thick layer of dead litter.

When such type of layer is allowed to develop, it tends to retain nutrients in a form that is not immediately available to roots. It also acts as a physical barrier for new shoots that grow from the soil surface towards the direction of light. In temperate grasslands, this layer acts as thermal insulation which slows down the spring warming of the soil. This has obvious implications for the grazing management of these systems.

Woody plants shade the grasses and dominate the vegetation, but are disadvantaged by the shortness of the growing season. However, in the absence of heavy mammalian grazing and regular fires, there are some trees and shrubs which can become established and grow vigorously. Therefore, in such situations, the grasslands are maintained by these natural or seminatural disturbances of fire and grazing. This in turn prevents the succession of the grassland vegetation toward tropical deciduous forest, savanna, scrubland, or temperate forest.

1.6. BIOLOGICAL PRODUCTIVITY

In a typical Canadian grassland, the annual productivity aboveground was found to be 6.4 metric tons per hectare when all the leaching and decomposition losses from dead shoots were considered.

The values which are found in other grasslands have greatly varied but most of the time they are considerably lower in comparison to this. This is partly because such losses have not been fully considered. The repeated harvesting (surrogate grazing) commonly yields 1.5 to 2.5 metric tons per hectare per year.

Frequently, grasslands are converted to cropland on which edible grains are grown. This allows for humans to take food from the grasslands directly instead of grazing animals feeding on the native grasses in a rangeland situation.

The increase in yields is said to be substantial. For example, as per the data compiled for the northern Great Plain of North America, one hectare of natural grassland, which is grazed by the cattle, can provide the equivalent of one- tenth of annual food energy requirements of one person, whereas the same area sown to wheat can provide requirements of four persons. However, it is important to point out that wheat yields cannot be sustained on an annual basis at their initially high levels without additional inputs of fertilizer.

In a similar way, sustained heavy grazing of natural grasslands can quickly result in the degradation of the vegetation as well as erosion of the soil. Particularly, this is likely to occur unnoticed in regions which lack fences and therefore do not exhibit fence line contrasts in range condition. In such fenceless grazing regions, degradation due to overgrazing may be difficult to differentiate from background changes because of long- or short-term weather variations.

1.7. TYPES OF GRASSLANDS

There are two main kinds of grasslands: tropical and temperate. Examples of temperate grasslands include Eurasian steppes, North American prairies, and Argentine pampas. Some of the tropical grasslands include the hot savannas of sub-Saharan Africa and northern Australia.

There is variation in the rainfall across grasslands from season to season and year to year. This ranges from 10 to 40 inches on an annual basis. There are temperatures below freezing in temperate grasslands to above 90 degrees Fahrenheit.There are variations in the height of vegetation on grasslands with the amount of rainfall. Some of the grasses might be under a foot tall, and some can grow as high as seven feet. There can be an extension in their roots up to three to six feet deep into the soil. The combination of underground biomass along with the moderate rainfall (most of the time, heavy rain can wash away nutrients) tends to make grassland soils very fertile as well as appealing to be used for agricultural purposes.

A large part of the North American prairielands has been converted into land for crops. This poses a threat to species which depend on those habitats, along with drinking water sources for people who are live somewhere around that.

1.7.1. Tropical Grassland

Tropical grassland biomes can also be referred to as savanna biomes. It is a terrestrial biome which features vast open spaces containing scattered small shrubs and trees. Savanna biomes or tropical grassland biomes tend to provide support to some of the most recognizable species of the world such as lions, hyenas, zebras, cheetahs, gazelles, giraffes, wildebeests, elephants, and warthogs.

Figure 1.3. Tropical Grassland.

Source: Image by Wikimedia Commons

Savannas of one sort or another cover almost half the surface of Africa (approximately five million square miles), generally central Africa, and large areas of Australia, South America, and India. Climate is considered to be the most important factor which helps in creating a savanna. Savannas are always found in warm or hot climates where the annual rainfall is from about 50.8 to 127 cm (20-50 inches) per year.

It is important that the rainfall is concentrated in six or eight months of the year, which is followed by a long period of drought when there can be an occurrence of fires. If the rain were well distributed throughout the year, there would be a number of such areas which would become tropical forests. There are savannas which are the result of climatic conditions and these are referred to as *climatic savannas*. The savannas which are formed by soil conditions and that are not wholly maintained by fire are referred to as *edaphic savannas*. The edaphic savannas can occur on hills, ridges, or where the soil is shallow, or in the valleys where clay soils become waterlogged during the wet weather.

There is a third type of savanna, which is referred to as *derived savanna*. This savanna is the result of people clearing forest land for the purpose of cultivation. Farmers tend to fell a tract of forest, burn the dead trees, and then plant crops in the ashes for the time period while the soil remains fertile. After that, the field is abandoned and, despite the fact that the forest trees may recolonize, grass grows on the bare ground (succession). This further becomes luxuriant enough to burn within a year or so.

In Africa, there is a heavy concentration of elephants that are in protected parkland. They have created a savanna by eating leaves and twigs, smashing the trunks, breaking off the branches, and stripping the bark of trees. Any dense woodland can be converted by the elephants into an open grassland in a small period of time. Then the annual fires maintain the area as a savanna.

The soil of the savanna is porous, and so, has rapid drainage of water. It contains only a thin layer of humus (the organic portion of the soil which is created by partial decomposition of plant or animal matter), which provides nutrients. Sometimes, savannas are classified as forests. The predominant vegetation contains grasses as well as forbs (which are small, broad-leaved plants that grow along with the grasses).

There are a number of different savannas that support different grasses because of the disparities in rainfall and soil conditions. Usually, only one or a few kinds of grass are more successful in comparison to the others in a particular area due to the support provided by savanna to such a large number of species which are competing for living space. For example, there are drier savannas such as those on the Serengeti plains or the Laikipia plateau of Kenya. Here, there are dominant grasses on the well-drained soils and those are the Rhodes grass and red oat grass.

In East African savannas, it can be said that the start grasses are dominant. In many western Uganda savannas, the lemon grasses are the common ones. Across the open landscape, the deciduous trees and shrubs are scattered.

There is one type of savanna which is common in southwestern Kenya, Tanzania, and Uganda. This savanna is referred to as *grouped-free grassland*. This has trees growing only on termite mounds, which is the intervening soil that is too thin or poorly drained in order to support the growth of trees at all. The seedlings are killed by the frequent fires and large grazing mammals. Therefore, the density of trees and shrubs is low.

Savannas receive an average annual rainfall of 76.2 - 101.6 cm (30 - 40 inches). However, there are certain savannas that can receive as little as 15.24 cm (6 inches) or as much as 25.4 cm (10 inches) of rain a year.

Savannas have both a dry as well as a rainy season. Seasonal fires play a very important role in the biodiversity of the savanna. In the month of October, there is a series of violent thunderstorms, which is followed by a strong drying wind which indicates the start of the dry season. Around the month of January is the height of the dry season where fires are prevalent. In the savannas, fires are often caused by the poachers who do want to clear away the dead grass in order to make it easier to see their prey.

The community is not devastated by the fires. A number of animals are killed by the fires, but to others the fire is a feast, like the birds which come to sites of fires in order to eat the grasshoppers, beetles, stick insects, mice, and lizards which are killed or driven out by the fire. For small creatures, a safe refuge is provided by underground holes and crevices. There are larger animals that are usually able to run fast enough so that they will escape the fire. Despite the fact that the dry stems and leaves are consumed by the fire, the deep roots of the grasses remain unharmed. Such roots, along with their starch reserves, are ready to send up new growth when the soil becomes much moister.

After a fire, scattered shrubs can subsist on the food reserves in their roots while they wait to venture above the soil again. Unlike grasses and shrubs, there are trees that survive a fire by retaining some moisture in all their above-ground parts throughout the dry season. Sometimes they tend to have a corky bark or semi-succulent trunk which is covered with the smooth resinous bar, both of which are fire-resistant.A fire can cover the earth with a fine layer of powdery black ash. During the month of March, there are violent thunderstorms that take place again, this time heralding the rainy season. When the rains come the savanna bunch grasses grow in a vigorous manner. In the time of 24 hours, some of the larger grasses tend to grow an inch or so. At this time the savannas tend to experience a surge of new life, such as the antelope birthing season. Mothers tend to have plenty of milk, along with plentiful grass to feed on. If the rains fail to come, then the calves die.

There are other animals which do not all occur in the same savanna. These animals include giraffes, buffaloes, zebras, lion, hyenas, leopards, and elephants. Also, there are some environmental concerns which are related to savannas such as poaching, overgrazing and clearing of the land for crops.

1.7.2. Temperate Grassland

Temperate grasslands have grasses as the dominant vegetation, with an absence of trees and large shrubs. There is variation in the temperature from summer to winter and less rainfall in comparison to that in savannas. The major examples are the veldts of South Africa, the pampas of Argentina and Uruguay, the puszta of Hungary, the plains and prairies of central North America, and the steppes of the former Soviet Union.

Temperate grasslands have hot summers, cold winters, and moderate rainfall. The height of grassland vegetation, along with the taller grasses in wetter regions, is influenced by the amount of rainfall taking place annually. Similar to the savanna, seasonal drought and occasional fires are very important to biodiversity. However, there are some effects that are not as dramatic in the temperate grasslands as they are in savannas.

Figure 1.4. The Puszta of Hungary.

Source: *Image by Needpix*

The soil of the temperate grasslands is said to be deep and dark, and is fertile in the upper layers. It is said to be nutrient-rich from the growth and decay of deep grass roots. Rotten roots tend to hold the soil together and therefore provide a food source for the living plants. Each species of grass grows best in a specific grassland environment determined by temperature, soil conditions, and rainfall.

Grazing by large mammals, seasonal drought, and occasional fires tends to prevent woody shrubs and trees from invading and further becoming the established. However, there are some trees such as cottonwoods, willows, and oaks which grow in the river valley.

There are also some non-woody plants (particularly a few hundred species of flowers) that tend to grow among the grasses.

There are a variety of species of grasses including purple needlegrass, buffalo grass, blue grama, and galleta. Some of the flowers inclide asters, coneflowers, blazing stars, sunflowers, goldenrods, psoraleas, clovers, and wild indigo.

In the temperate grasslands the precipitation usually occurs in the late spring and early summer.

The annual average is about 50.8 to 88.9 cm (20 - 35 inches). Summer temperatures can be well over the temperature of 38 °C (100 °F), whereas the winter temperatures can be as low as - 40 °C (- 40 °F).

The fauna does not occur in the same temperate grassland. The fauna includes gazelles, zebras, wild horses, rhinoceroses, wolves, lions, jackrabbits, prairie dogs, mice, foxes, coyotes, blackbirds, skunks, badgers, meadowlarks, grouses, sparrows, quails, owls, hawks, grasshoppers, snakes, spiders, and leafhoppers.

There are also environmental concerns related to temperate grasslands. There are few natural prairie regions that remain because most have been turned into farms or grazing land.

The reason is that they are flat, treeless, covered with grass, and have rich soil.

There can be a further division that can be done in the temperate grasslands. Prairies are grasslands that have tall grasses whereas steppes are considered to be the grasslands that have short grasses. Prairie and steppes are much similar but the information that has been mentioned above particularly pertains to prairies.

Steppes are dry areas of grassland that have hot summers and cold winters. They tend to receive 25.4-50.8 cm (10-20 inches) of rainfall annually. Steppes tend to occur in the interior of North America and in Europe. The plants that grow in steppes are usually greater than 1 foot tall.

Blue grama grass, buffalo grass, sagebrush, cacti, spear grass, and small relatives of the sunflower are all included in the plants growing in steppes. Badgers, owls, hawks and snakes are some of the fauna that live in steppes. In today's world, people use steppes for grazing livestock and also for growing wheat and other crops.

Some steppes have been harmed because of overgrazing, plowing, and excess salts that are left behind by irrigation waters. After plowing, the strong winds blow loose soil from the ground, especially during droughts. This is the reason for the dust storms of the Great Plains in the US.

1.8. CHARACTERISTICS OF THE GRASSLAND

There are different types of grasslands which share similar characteristics. Savannas have scattered trees and predominate in some specific parts of Africa, South America, Australia and Asia.

The temperate grasslands are majorly devoid of trees, they endure broader temperature extremes, and receive less rainfall in comparison to savannas.

There are two types of temperate grasslands which include steppes and prairies. The steppes contain shorter grasses whereas prairies have taller grasses due to higher precipitation.

Throughout the region of North America and Europe, both the types of temperate grasslands can be found.

1.8.1. Precipitation in the Grasslands

Around 25 percent of the land surface of Earth is grassland, primarily in regions with limited rainfall to prevent forest growth. This limited rainfall is due to nearby mountain ranges which cause rain shadows over adjacent open range lands.

Typically, grasslands have limited, unpredictable rainfall and droughts are common. Deserts will form where there is not enough rainfall. On average savannas receive roughly 76 to 101 cm (30 - 40 inches) of rain per year. Steppes receive an average of 25 to 51 cm (10 20 inches) annually. Prairies are the intermediate in between savannas and steppes with 51 to 89 cm (20 - 35 inches) annually.

Figure 1.5. Countryside Steppes Mongolia.

Source: Image by Pixabay

1.8.2. Temperature in the Grasslands

There can be variations in the temperatures and it can be much more in the temperate grasslands in comparison to savannas. Savannas are in warm climates along with average annual temperatures which only vary between 21-26 °C (70-78 °F). Typically, there are only two seasons namely a wet season and a dry season.

There are hot summers in temperate grasslands where the temperature can exceed 38 °C (100 °F) and cold winters that see temperatures below -40 °C (-40 °F).

1.8.3. Fire in the Grasslands

Fires are an important grassland characteristic. They promote the growth of native grasses, but can limit the growth of trees. There are deeper root systems in the native grasses that can survive fires, but invasive plants have shallower roots and so, succumb to fires.

The lack of seasonal fires has threatened the health of worldwide grasslands and development has curtailed the number and extent of grassland fires. Only around 5 percent of the grasslands present in the world are being protected and maintained, therefore they remain the most endangered biome in the world.

1.8.4. Flora and Fauna

Savannas are considered as a home for some of the largest mammals on the planet which include elephants, giraffes, rhinos, lions and zebras. Temperate grasslands are home to large animals, specifically bison and horses, medium-sized mammals such as deer, antelope and coyotes and also, some small mammals which include mice and jackrabbits.

The type of grasses which grow depend upon the amount of rainfall. Most of the time, shorter steppe grasses contain buffalo grass and savanna grasses will contain taller grasses such as bluestem and rye.

1.9. GRASSLAND HABITATS

Grassland habitats are an increasingly rare sight in New Hampshire. These open areas of fields and wildflowers are used by more than 70 species of wildlife in order to meet their needs for food, cover or breeding.

Figure 1.6. New Hampshire, England.

Source: Image by Pixabay

A number of grasslands present today are the result of the land clearing, therefore they require maintenance. If these habitats are left alone then these habitats will grow shrubs and small trees, which ultimately reverts the habitat to forest. It is important to learn to recognize the habitat value of grasslands as well as discover what can be done in order to maintain and conserve these special habitats.

The vegetation which grows in grassland habitats can be tall (over four feet), short (less than 6 inches) or a combination of both. The height of the vegetation plays an important role in order to determine which wildlife species will use the habitat.

Currently, there are a number of non-native grasses which grow in grasslands. These are introduced by humans for agricultural uses. These grasses include timothy, orchard grass, Kentucky bluegrass, and perennial ryegrass. There are two native kinds of grass, big bluestem and little bluestem, and native wildflowers namely goldenrod and aster, which are common in our grasslands today.

1.10. TYPES OF GRASSLAND HABITATS IN NEW HAMPSHIRE

Agricultural Fields

The most common grassland habitats found in New Hampshire are agricultural fields such as hayfields, pastures, and fallow fields. Vegetation contains a mixture of grass species or a combination of grasses, wildflowers, and sedges.

Other Grasslands

Airports, capped landfills, military installations, and wet meadows can also function as grassland wildlife habitat if they support similar vegetation. A number of grassland wildlife species use croplands as potential grasslands, as they may be easily converted to grow grass if the crop farming practices are abandoned.

Figure 1.7. Wet Meadow.

Source: Image by Flickr

1.11. HISTORICAL CHANGES IN GRASSLAND HABITATS

Historically, Native Americans and beavers were the key forces responsible for forming and maintaining grassland habitats in New England. Native Americans created grasslands when they burned the land for agriculture and to improve the forage for game species like white-tailed deer. Simultaneously, ponds above the abandoned beaver dams developed into grassy meadows after the water drained and the nutrient-rich soil was exposed to the sunlight.

In the past few years, fire suppression and limits to where beavers are permitted to construct dams have meant that grasslands are only allowed for agricultural areas. The peak of agricultural clearing in the state took place in the mid-1800s. From that point of time, New England has been losing grassland habitats which have reverted into forest. Along with their well-drained soils, tree-less fields, and a huge amount of road frontage, agricultural lands provide attractive sites for development.

Presently, most of the grasslands in New Hampshire need proper maintenance by humans. If these habitats are left alone, small trees and shrubs will grow, reverting ultimately into forest.

1.11.1. Where are New Hampshire's Grasslands?

Development and succession of natural forests have combined to decrease the grassland habitats in New Hampshire to the point that grasslands cover just around 4% of the total landscape.

Large grassland habitats that are greater than 25 acres in size are still present in almost every county in New Hampshire, with the major concentration in Grafton County (almost 20% of remaining grasslands), Merrimack County (13% grassland), and Coos County (12 % grassland). A certain level of conservation restriction protects around 8% of New Hampshire's large grassland habitats.

1.12. CLIMATE VULNERABILITIES FOR GRASSLANDS

- Usually, such habitats are considered to be resistant to climate change as long as the management continues.
- Increased amount of diversity and plenty of invasive mainly plants.

- The potential for wildlife loss value to such habitats from increased and modified management resulting from a long growing season and also an increased demand for the biomass fuels.

1.13. AGRICULTURAL PRACTICES & BIRD NESTING

Without the effective work of farmers and other landowners, most of the grasslands would revert to forest quickly. However, the timing of mowing can eventually affect the ability of a field to provide habitat for grassland-nesting birds and some other wildlife.

Farmers who grow high-quality forage for livestock generally mow their fields around two to three times during the summer period. Therefore, one such mowing event usually takes place between May and mid-July. This is considered to be the time that corresponds with the nesting season for most of the grassland-nesting birds. Mowing during this period can eggs and nests, kill fledglings, or cause adult birds to abandon their nests.

1.14. GRASSLAND-NESTING BIRDS

Bird species that depend on the grassland have been reduced along with their habitats, much faster than any other group of birds in New England. Most of the grassland-nesting birds are found to be 'area-sensitive'. This means that they will not nest in fields that are smaller than a particular size.

Table 1.1. Minimum field size required for the grassland bird species

Grassland Bird Species	Required Minimum Field Size	Preferred Vegetation Height in Fields
Bobolink	5+ Acres	Dense grass taller than 3 feet
Eastern meadowlark	15+ Acres	Dense grass and wildflowers taller than 3 feet
Savannah sparrow	20+ Acres	Short and tall vegetation
Grasshopper sparrow	30+ Acres	Short, sparse grass Listed as a state-threatened species
Northern harrier	30+ Acres	Short grass fields and wet meadows State-endangered species
Upland sandpiper	150+ Acres	Short, spare grass Very rare, state-endangered species

Source: https://extension.unh.edu/resource/grassland-habitats

1.15. GRASSLAND MANAGEMENT

Grassland management is the branch of agriculture involved in the production of green fodder, hay, and raw materials for grass meal and certain other fodders on sown and natural hayfields and pastures. It is considered the manipulation of natural vegetation so as to attain some predetermined goals. Grasslands are usually managed to enhance productivity and increase benefits for human use.

It is crucial to understand the relationship between environmental factors and the species that make up the grassland ecosystem so that proper management techniques can be used.

Contrary to forests and bogs where human intervention has a negative impact most of the time, grassland ecosystems in the Baltic countries are a result of the interaction between humans and nature. Thousands of years ago, in the dense woods of the region, the first tribes used to slash and burn agriculture which formed space for grassland species.

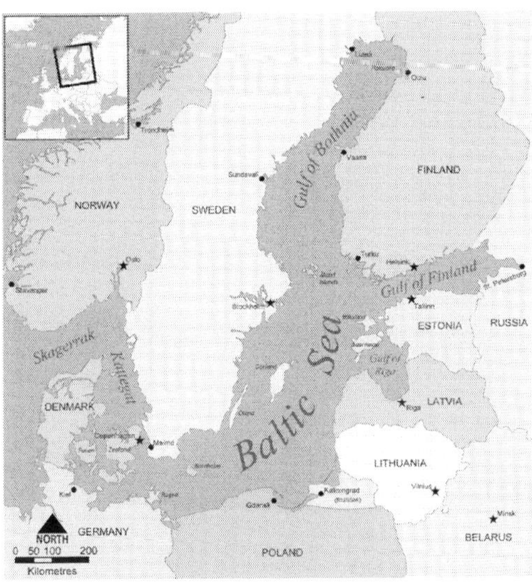

Figure 1.8. Baltic Countries.

Source: Image by Wikimedia Commons

There is extensive management done by humans which is considered an important part of the ecosystem. This management created species-rich meadows and pastures, which have survived for a number of centuries. As agriculture became more intensive, unfavorable conditions began to arise.

There was a decrease in the extent of the areas having traditional grassland management methods. A number of grasslands were then converted into cropland or sawn grasslands having just a few species.

Marginal grasslands were abandoned which in turn resulted in the overgrowth of shrubs and trees, converting the ecosystem to forest. Therefore, it is important to ensure proper management conditions for grasslands to maintain their natural state.

Grassland management is concerned with the production of fodder on both meadows as well as some other types of forage lands such as bogs, steppes, and deserts. It plays a vital role in creating a stable fodder supply for public livestock production. Each year, hayfield and pastures generate around a third of all fodder that is grown in the USSR.

There are some main goals of grassland management and they include:

- to develop highly productive planted lands,
- to improve natural and sown hayfields and pastures, and
- to ensure the most efficient use of all of these lands.

Considering the point of view of the organization as well as economics, grassland management is closely related to field cropping and animal husbandry.

The history of grassland management can be divided into four periods. During the very first period, there was primitive use of forage lands. The cattle grazed freely throughout the whole year. In the winter, they are found to fodder under the snow. The animals were driven from one seasonal pasture to another seasonal pasture.

During the second period (around the 11th and 12th centuries), grasslands were used as pastures and hayfields. The cattle grazed and hay was laid in both as the main fodder for the winter. The cattle were stabled and kept as a reserve for the period when the animals were maintained in distant pastures.

In Russia, when the large quantities of agricultural produce became important for the cities, the beginning of the third period more or less coincided with the emancipation of the serfs and the development of industrial capitalism. There was organization of the experiment stations, bases and fields, and demonstration plots which was done in order to develop the methods for improving the natural forage lands. However, these efforts were of practical value only for large estates.

The fourth period, referred to as the modern period, started after the Great October Socialist Revolution. In the year 1920 land reclamation subsections were organized in provincial and district land agencies. Furthermore, these subsections were staffed with hydraulic engineers, land reclamation engineers, and grasslands specialists.

The land improvement systems were developed by these specialists on natural forage lands. These specialists improved them superficially, developed drained lands, and sowed hayfields and pastures. A great deal of work was done on the floodplains of the Dubna, Iakhroma, Viatka, and Moscow rivers.

1.16. CONCLUSION

Grasslands occupy approximately 70 percent of the world's agricultural area. This is approximately half of the ice-free land area of the world. These are especially important in areas where people are the most food insecure. There is significant potential for carbon (C) sequestration and emission reductions, but still, they are not included in international agreements to reduce greenhouse gas (GHG) emissions.

Grasslands are considered as the most widespread of all the major vegetation types of the world. However, human activities have altered most of the vegetation types of the world. Creating artificial grasslands of cereal crops requires some form of repetitious, unnatural disturbance including cultivation, heavy grazing, burning, or mowing.

Grassland habitats rarely exist in New Hampshire. There are more than 70 species of wildlife who use the open areas of fields and wildflowers to meet their needs for food, cover, and/or breeding. The discussions mostly concentrate on the natural and nearly natural grasslands.

Grassland management is the management and production of fodder, both on meadows and on other types of forage lands (bogs, steppes, and deserts). The grassland management has an important role to play in the creation of a stable fodder supply for public livestock production.

REFERENCES

1. Extension.unh.edu. (2018). *Grassland Habitats | UNH Extension*. [online] Available at: https://extension.unh.edu/resource/grassland-habitats [Accessed 24 Feb. 2020].
2. KONIUSHKOV, N. (2010). *Grassland Management*. [online] TheFreeDictionary.com. Available at: https://encyclopedia2.thefreedictionary.com/Grassland+Management [Accessed 24 Feb. 2020].
3. M.B. Smith, J. (1999). *grassland | Definition, Animals, Plants, Types, & Facts*. [online] Encyclopedia Britannica. Available at: https://www.britannica.com/science/grassland [Accessed 24 Feb. 2020].
4. Madaan, S. (n.d.). *Tropical Grassland (Savanna Biome): Climate, Precipitation, Location, Plants, | Earth Eclipse*. [online] Earth Eclipse. Available at: https://www.eartheclipse.com/ecosystem/tropical-grassland-savanna-biome.html [Accessed 24 Feb. 2020].
5. NUNEZ, C. (2019). *Grasslands*. [online] National Geographic. Available at: https://www.nationalgeographic.com/environment/habitats/grasslands/ [Accessed 24 Feb. 2020].
6. Peterson, J. (n.d.). *Characteristics of the Grassland*. [online] Sciencing.com. Available at: https://sciencing.com/characteristics-grassland-23689.html [Accessed 24 Feb. 2020].
7. Ucmp.berkeley.edu. (n.d.). *The grassland biome*. [online] Available at: https://ucmp.berkeley.edu/exhibits/biomes/grasslands.php [Accessed 24 Feb. 2020].
8. Vivagrass.eu. (n.d.). *Grassland management | LIFE Viva Grass*. [online] Available at: https://vivagrass.eu/grasslands/ [Accessed 24 Feb. 2020].
9. World Wildlife Fund. (n.d.). *Grasslands | Habitats | WWF*. [online] Available at: https://www.worldwildlife.org/habitats/grasslands [Accessed 24 Feb. 2020].

2

Production and Conservation of Grasslands

CONTENTS

This chapter is about the Grassland production process, their conservation methods, causes of loss of nutrients, different types of conservation practices. This chapter focuses on the production of vegetative grassland in an effective manner. With that, this chapter also explains the importance of maintaining grassland production.

With all these things, this chapter also focuses on the threats that are responsible for effecting those grasslands. These threats are affecting the vegetative grassland. these threats are basically responsible for making the vegetative grasslands into barren lands or into desserts. So, it is important to maintain those vegetative grasslands to avoid making them barren.

2.1. INTRODUCTION

Grasslands are areas which are dominated by grasses. As grassland areas are very important for a healthy environment, maintaining and improving production of these grasslands is also very important.

Grassland biomes occur naturally worldwide. In North America alone there are about 1.5 million square kilometers of grassland. There are three types of grassland that are commonly found in this area: tall, mixed, and short grass prairies. Each of these grasslands provides valuable ecosystem services.

There are ecological services that are provided by the grasslands including habitats for wild species, carbon storage, forage for livestock, and habitats for pollinators. The degradation of grasslands is basically affected by their ability to function properly. This hinders the ability of the grassland to provide the full suite of ecological services. There are many causes of degradation that may vary from region to region.

Figure 2.1. Image of grassland Biome.

Source: Image by Pxhere

It is considered that much of the tall-grass prairie has been plowed for crop production. Nowadays, these grassland areas are being degraded by cutting down all the vegetation for the development of urban areas. These developments have created more area for human habitation, but they have negatively affected the natural habitat.

2.2. PRODUCTION OF GRASSLAND

Grasslands are important to land that is used as an agricultural area. Grasslands play an important role in feeding herbivores and ruminants. Grassland provides important ecosystem services that include erosion control, water purification, and water management.

Grassland supports biodiversity and cultural services. These services include recreational values and stock of carbon (Smit et al., 2008). In terms of management, yield, and biodiversity values, grasslands are considered as diverse. Their range starts from the seminatural grassland with low yields and biodiversity values in high amounts to fertilize mono -cultural grassland.

Figure 2.2. Representation of importance of Grassland.

Source: Image by Needpix.com

It is said that most of the grass in the EU originates from intensively managed grassland and gets stimulated by fertilizer application. The low yields are considered to be the result of the high natural value of the grassland. These include mountain summer gazing areas, semi-natural grasslands, and other areas that are used for extensive grazing.

The agriculture ecosystem collects solar energy and stores it through photosynthesis. This energy is stored in the form of carbohydrates, proteins, and lipids. There are four processes that can be used to describe the use of accumulation of plant biomass and the interception of solar radiation (Sinclair and Weiss, 2010).

2.2.1. Interception of the Photosynthetically Active Radiation

It is estimated that half of the energy in solar radiation is photosynthetically active radiation (PAR). When the fraction of the PAR gets intercepted by the canopy which is basically dependent on the extent of the leaf area index (LAI).

Leaf area index is the ratio of the leaf surface area on one side, per unit ground surface area. The LAI is roughly 4 to 6 in the case of many established crops in the agriculture system.

The grass swards, at the time of sowing the value of LAI may range zero but in case of full grain crop the LAI value that may range up to nine (Simon and Lemaire, 1987). When the value of LAI is 1.0 then the PAR gets intercepted by approximately 50%, when the value of LAI is 3.3. then the PAR gets intercepted approximately 90% and when the value of LAI is 4.3 then the PAR gets intercepted by 95%.

When an adequate supply of water and nutrients is provided, these factors can help in promoting a rapid development in the value of LAI. There are other factors that are bringing a rapid development of the value of LAI and those are vigorously growing genotype, adequate plant density, and arrangement of the plants in an optimal spatial arrangement.

2.2.2. Use Efficiency in Radiation

Leaf pigments absorb intercepted PAR which is mainly chlorophyll and is used to produce the chemical energy in the plant. The energy is said to transfer in several steps which include energy- rich intermediate adenosine triphosphate (ATP) and nicotinamide adenine dinucleotide phosphate (NADPH). ATP and NADPH, in turn, provide the energy that can be used to change into simple organic molecules which gets assimilated by the carbon dioxide (CO_2).

The plants that are identified as C3 species produce three-carbon sugar which works as building blocks for other components. Whereas, another group of the plant which is identified as C4 species, the initial capturing of the CO_2 takes place in the four-carbon organic compound. These acids are then transported to specialized cells. In these cells carbon dioxide is released and is processed again as in C3 plants. C3 grassland is commonly considered as a temperate grass, whereas C4 grassland as a tropical grass.

Figure 2.3. A figure of the effect of chlorophyll on plants.

Source: Image by Wikipedia

The amount of assimilated carbon dioxide is considered as a function of the amount of PAR energy that is stored in ATP and NADPH. Therefore, the amount of dry matter which is produced is directly related to the amount of absorbed radiation.

The accumulated plant mass per unit of intercepted radiation is defined as the radiation use efficiency, which is symbolically known as RUE. Photosynthetic capacity of the individual leaves, a high capacity that leads to high dry matter accumulation but with a diminishing slope, are some of the important factors for RUE.

The value of RUE further depends on the composition of the plant. Carbohydrates require less energy than proteins or lipids. Therefore, plants that contain high protein or lipid content are said to contain low RUEs.

2.2.3. Length of Growing Season

The total mass that gets accumulated in the plants during a growing season is derived from the sum of the daily biomass growth values, which is the product of RUE and intercepted radiation. The regular values of radiation are considered to be dependent on the location which includes latitude, elevation, and topology. It is also said to be dependent on short-term weather variability.

In the case of grassland, the amount of intercepted radiation shows more change. These changes are due to the harvest sequence throughout the season. There should be sufficient regrowth after cutting and grazing in order to achieve LAI to again intercept a high amount of radiation. For much of the growing season, the RUE is said to relatively stable. Therefore, there is a decrease during the seed growth of the later stages. When there is a transfer of nitrogen from a vegetative component like from the leaves, which will lead to a sequence of events that can result in decreasing the LAI and RUE during reproduction.

During the growing season, some factors such as temperature are considered as the most critical factors for the length. For the proper growth of the crop, an average temperature of 50°C is often used as an important rule.

2.2.4. Partitioning of Plant Mass

The total yield is determined by the fraction of the total plant mass that is to be harvested. The harvestable biomass when divided by the total above-ground biomass is used to define the harvest index (HI) of the grass. The HI of the forage grasses depends on the residual biomass, which is left after cutting or gazing. The value of HI is normally around 0.7 to 0.8 (Larcher, 2003; Jing et al., 2011).

Each of the processes which are explained above are sensitive to environmental stress which result in an inability to attain the maximum yields. The most common explanation for the lower yields is inadequate water availability and nutrient supply. But there are other factors that are also relevant such as extreme temperature, competition with weeds, or pathogens.

2.3. METHODS FOR THE MEASUREMENT OF PRODUCTION OF GRASSLAND

2.3.1. Unit of Yield

The biomass of grassland vegetation is referred to as the above-ground herbaceous material. This can be commonly referred to as a dry matter (DM) yield. Vegetation mass can be also expressed as an organic matter (OM) instead of referring it to dry matter yield. OM yield has the advantage that it is a true herbage yield that is without the soil contamination. This is said occurs in herbage samples which result from mechanical cutting, racking up of the grass, or rain splash. In the case of intensive animal production systems, grassland yield is said to initially sampled as a dry matter or organic matter. This can be often expressed in terms of feeding units that are based on the net energy value.

2.3.2. Destructive Measurement

In the country like Europe, destructive measurements are commonly applied to the grassland experiments. The most common and simplest devices that are used for this purpose are hand-operated tools such as scissors, secateurs, knives, shears, sickles, and scythes.

When a material is to be divided into species or groups of species then the small -handed tools are taken as a useful tool for small plots. In the case of short to medium- tall swards, engine- driven reciprocating cutter bar mowers and lawnmowers with a catcher can be used for this purpose. Commonly, a Danish harvester or Haldrup that is used is said to consider a cutting width of 1.5 m with the adjustable stubble height. A sample of cut material is taken by hand or it can be taken automatically by using new models. The total fresh weight is automatically recorded for each plot. It is important with any type of cutting that implements that the cutting height above the ground level can be also controlled. For the near ground level, the use of hand- held shears or secateurs for cutting purposes is used.

Figure 2.4. People plucking weeds and poppy plants to have good vegetative grassland using destructive method

Source: Image by Wikimedia common

Therefore, this process of cutting the ground level will affect the regrowth and sampling area as well. It is considered that sampling from those areas should be avoided in the future. Cutting height will vary depending on the type of grassland. The cutting in those areas will range from 1 cm which is close to grazed pastures to 10 to 20 cm in tall swards.

In many experiments, the grassland is cut at 4 to 5 cm. Lower cutting heights and mechanized equipment's can even suck some extraneous materials. These materials can be detached litters, twigs, dry feces and gravels.

2.3.3. Non-Destructive Measurement

Destructive sampling is said to require high inputs of labor or equipment. Therefore, this method can be costly and may lead to insufficient sample numbers, which in turn results in low precision. Destructive sampling may also prevent some of the measuring changes of the sward in the sampling area. The time at which the material is removed by cutting in the small grazed plots may be significant in the proportion of the feed available.

For these reasons, a non-destructive sampling technique is being developed. This technique can be grouped into visual estimation, density measurement and height, and the measurement of the non- vegetative attributes can also be related to the DM yield. Visual assessment may be carried out by experienced operators.

These operators are very similar to the type of posture that is under consideration. This may be helpful to estimate the dry matter yield within the circa one ton per hectare.

However, this procedure is of limited value to research without any calibration. In grassland research, the comparative yield method is being widely used. This method is used as a regression equation which is derived from standards covering a range of dry matter yields. They are used to calculate the dry matter yields which are based on the estimates from operators.

The standing of the biomass of an area of grassland is related to the density and the height of the individual components. Height and density measurement of the sward is integrated using a weighted disc, raising plate, pasture disc, or drop disc. They are said to consist of a round or square disc made up of light metal or plastic foam. This material is of given weight that can slide along a central rod. This is then lowered or dropped from a fixed height onto the sward.

Figure 2.5. The image representing non-destructive method for maintaining vegetative grassland.

Source: Image by National Park Service

In order to estimate the yield, the capacitance meter device is used. This process or method is based on a signal that is produced by an oscillator; these signals are produced in an electrical circuit. Under measuring head change, these changes as a capacitance. Material that contains high capacitance is herbage mass, whereas air and wood contain low capacitance.

The indirect method that is used to measure the dry matter yield is the difference in capacitance between a quadrat on bare ground and on a grass sward. Since 1956, capacitance meters have been used. Although there is an improved version that has developed. But their performance still leaves much to be desired. This is an exception under certain or special circumstances.

The method of non-destructive is often used as a double sampling technique, that is, there are two overlapping methods that are used. These two overlapping methods are, one is an accurate determination which involves DM yield in a few standards.

Whereas, others involve visual estimate of height or capacitance reading of herbage in many samples which basically includes standards. There are some regression equations that provide the calibration of the technique. These equations are between the estimates non -yield parameters and DM yield of the standards.

Therefore, the non-destructive technique still requires some sample cutting, but the amount included is a small quantity. The cutting can be restricted to an area of the same sward, that can be outside the measurement or treatment area.

2.3.4. Remote Sensing

Remote sensing is considered as an important tool that is used for monitoring the land cover classes that include grasslands. The relevance of remote sensing as a source of information for grasslands is being conditioned by the sensor characteristic with spatial resolution, temporal resolution, and spectral resolution.

In terms of spectral signatures, the types of description of habitat and types of distinction of grassland can be improved by better understanding. Their special and temporal variations can be also improved with that. The description of the textural features like tone, texture, structure, and pattern of different types of satellite imagery that are depicting the grasslands.

In order to identify the spatially covered land with class grassland and types of grassland in some specific cases can be done by remote sensing. There are many different local and regional grasslands that are classified in the scheme such as floristic, habitat, climate, and management. In most cases the floristic composition plays an important role and is not that easy to distinguish from satellite imagery.

In order to identify the grassland parameters, remote sensing is mostly included. This includes Leaf Area Index, a fraction of cover, the gap in the fraction of the soil like bare land, the canopy shade, biomass content, and canopy coverage. Methods that are used for assessing the plant canopy biophysical parameters such as green leaf area and biomass are the rapid and non-destructive methods that provide a remote measurement of canopy spectral reflectance.

Red and near-infrared reflectance are found to correlate with the amount and duration of the green leaves. But there are some factors that are also dependent of the site and data set effects. These factors are empirical relationships that is between canopy biophysical parameters and spectral reflectance.

The European Commission is needed to provide timely information on the agriculture production that is to be expected in the current season for the implementation of the common Agriculture policy. This was the important concern of the MARS project which is Monitoring Agriculture by Remote Sensing of JRC.

It was recognized that remote sensing was derived by the indicators that play a major role in forecasting the crop yield in the country Europe. This was done despite focusing initially on the use of remote sensing techniques that are used for crop yield forecasting within the MARS project.

The approach that is used for quantitative yield prediction within the MARS project. This gets gradually shifted towards an agrometeorological approach. Remote sensing was derived from indicators of the growing season that were merely used as a qualitative descriptor.

The short remote sensing can be attributed to the following:

1. Identification of the land that is covered with the grassland and in some cases the specific type of grassland. There is the availability of many different local and regional grassland classification schemes like floristic, climate, habitat, use, and management. In most of the cases, floristic composition plays and important role. But on the other hand, that is not easy to distinguish from the satellite imagery.
2.. Identification of the grassland parameters. These parameters include other LAI, canopy, fraction cover, gap fraction soil, soil moisture, biomass content, and canopy coverage. There are some biophysical parameters that can be used to retrieve from the GMES data source, such as GEOLAND-2. These GMES data sources are amongst others:
    a) In order to quantify the green vegetation, there are some classical parameters like LAI and FAPAR. They are strongly correlated with the fCover. It is used as a direct input into the grass vegetation density product, but the relation between LAI and fCover is far from linear.
    b) fCover: the useful parameter that is used for many environmental and climate related applications is Fractional green vegetation cover (FVC). When it is compared to the previously used NDVI, it is considered to have several strong advantages. These advantages are absolute parameters that can be included as sensor-independent, robustness to thin the clouds, fully scalable at different spatial resolutions.
    c) Canopy Shade Factor (CFS): This canopy shade factor parameter allows to characterize the amount of level of shadows that are self-casted on the canopies. This parameter allows many conditions to discriminate against

the rough canopies from the flat homogeneous canopies which include crops and grasslands.

d) fSoil: fSoil quantifies the gap fraction of the soil in the image. It relies on the capacity that can discriminate a third contributor. These contributors are brown or non-photosynthetic vegetation (NPV). fSoil can be useful in identifying intensive agriculture practices with bare soil events.

Remote sensing also plays an important role in grassland transpiration, fluxes in grassland emissions, phenology and grassland dynamics, grassland productivity, chlorophyll, grassland albedo, and water content with vegetation condition and structure.

2.3.5. Modeling

Both mechanical and empirical models consist of a wide variety in modeling the approaches of grassland production. In order to estimate the grassland production, there are some purposes that are used, these purposes are farming system research and collection of statistical data on a regional level and feed balances.

The mechanistic models are often based on the growth, senescence, litter, and standing biomass to predict grassland production. Grassland production also uses data on incoming radiation, soil moisture, temperature, day length, and altitude.

From the field of experiments, empirical models are derived from or are calibrated with the data. These field of experiments are experiments in which the effect of nitrogen application have been tested. In order to derive the equation, regression analysis is used. For example, growth curves for the potential dry matter yield with any limitation like water.

There are some fields that can be used to estimate the grassland yields using data on milk production, feed composition, energy requirement, and the maize yields in a farm, region or country. The grassland yield is considered as the output for the balance calculation.

The main advantage that is used for calculating the feed balance model. This method is a fairly simple method which is based on the statistics. This method takes an account of farm management practices, fertilizers use, feed and forage, livestock, and various soil production. Grass consumption is taken as a part of the feed that is consumed and is also a part of the minerals which end up in manure. On the other hand, it can be taken as an output of the soil surface balance. This implies that the inaccuracies in the calculation of the grass consumption cannot affect the soil surface balance.

2.4. MEASURING NUTRIENT CONTENT IN GRASSES

2.4.1. Destructive Measurement

Sampling: It is important to take a representative sample of the grass in order to analize the N and P contents. This is because of the spatial variability of the content may be high in the field. Natural or extensively managed grasslands have the highest spatial variability with the large diversity of the grass species. Grazed grassland with locally high content of N and P which related to the urine patches and pats and farms fields.

In well-managed plots that are relatively smaller in size, the spatial variability will be much smaller which are said to be in a controlled field of experiment. In the case of cut grasslands, the samples can be taken from the cut grass. Whereas, in the case of gazed grassland, the samples should be taken at different intervals and from different sites of fields during the gazing period. This is done in order to obtain an average value of the N and P contents of the grass that is consumed by livestock.

In the case of natural or extensively managed grasslands, samples should be taken from representatives of the species that are found in the specific feild. Individual plant scan is also sampled, beside harvesting the parts of the grassland area by mowing. In the year 1998, Kara suggested for hay, pasture, and forage grass in order to take 40-50 plants and to sample the fourth uppermost leaf blades.

The third way of sampling that is used does not include the grass in the field in the sample, but instead, grass in the silage pit or stored hay is sampled. It should be noted that at the time of sampling, silage and storage of hay will lose some amount of N. Therefore, samples should be collected from different places in the silage and stored hay to be a representative sample for the whole harvest.

Sampling, Handling and Preparation: In order to minimize the respiration and decomposition of the plant samples, these plant samples should be dried as soon as possible and stored in cool conditions. Fresh plants tissues should be placed in open, clean paper bags and should be partially airdried if possible. Plant tissues can be also kept in a cool environment during shipment to the labs (Kara, 1998).

Until the tissues are air-dried or the bag and content are kept in a cool temperature, fresh plant tissues should not be placed in a closed plastic bag. In order to dry the tissues, temperatures ranging from 60 to 80°C are used in a well-ventilated drying oven, and sometimes kept there for almost 24 hours. The enzymes that are present in the plant tissue will become inactive when comes under temperature above 60°C. When the temperature lies above 80°C, thermal decomposition and reduction in the dry weight will occur.

Only in case of room temperature air drying, the sample may not get stabilize and will prevent the enzymatic decomposition. Therefore, after taking the sample, the sample should be dried as soon as possible. After drying the sample is finely grounded. This allows the sample to pass through a sieve of 1mm size.

This sieving is done to obtain a homogeneous sample from which the sub-samples are taken. After the sample particle size gets reduced and homogenized, the sample should be stored in a cool and dry place in a tightly closed flask or bag in order to protect them against the sunlight. Dry samples can be kept or stored for almost 10 years (Houba et al, 1995).

Determination of Dry Matter Content: The dry matter content of the plant sample can be determined by the gravimetric loss of water. In this process, the material is dried at 105°C for almost 2 hours. The dry matter is often used to correct the concentration of the sample which is dried at 70 to 80°C to an absolute dry matter basis.

Drying the sample at a temperature of 105°C changes the chemical composition of the plant material. The sample is dried at this high temperature because the samples are not used for any chemical analysis.

Organic Matter Destruction and Chemical Analysis: By using decomposition or destruction of the organic matter, the plant tissue samples which are previously dried, ground, and weighed are prepared for the element analysis. The two most used methods of organic matter destruction are dry ashing which includes high- temperature combustion and wet ashing which includes acid digestions. Both methods are based on the oxidation of organic matter, which takes place by using heat and acid (Kara, 1998; Temminghoff et al, 2000). Dry ashing is conducted in a muffle furnace at a temperature range of 500 to 550°C for about 4 to 8 hours. Once the process of ashing ends, the vessel is removed from the muffle furnace, then it is cooled and the ash is dissolved in the acid.

Through the use of both heat and acids, wet digestion involves the destruction of the organic matter. In order to maintain the temperature of about 80 to 125°C, the hot plates or digestion blocks are frequently used for this process. In some of the cases, microwave heating is used which involves microwave digestion.

The most common method that is used to determine the NH4 and Protein N in the plant tissue with a well-known wet oxidation method is known as Kjeldahl method. This method is based on the wet oxidation of organic matter that uses sulphuric acid and digestion catalyst. The Kjeldahl procedure is said to have several variances that include mainly micro and macro. They are based primarily on the sample size and required apparatus.

2.4.2. Non-Destructive Measurement

A chlorophyll meter is an easy way to predict the crop N status. A strong positive relationship is maintained between the leaf chlorophyll content and the leaf N concentration. The hand-held chlorophyll meter permits an *in situ* rapid and non-destructive determination. This determination of leaf chlorophyll content takes place by measuring the leaf transmittance.

However, there are certain factors that are affecting the chlorophyll meter reading by crop cultivar. These factors are the stage of growth, soil moisture status, and nutrients other than N. The need of N fertilizer application can be detected by using the chlorophyll meter. Chlorophyll meters are considered to contain the greatest sensitivity. This sensitivity is deficient in the range of the N nutrition. Monochromatic light is directed at the plant tissue sample directly by using the near-infrared reflectance spectrometry (NIRS). In this method, the diffuse light is deflected from the sample and detected by lead sulfide detectors.

The instrument is most often used to determine the content of the protein. It is necessary to determine the same parameters which are via wet chemistry on a large population of samples, which is used to calibrate the NIR instrument. The major advantage of NIR is that the analysis is non-destructive, comparatively simple, and very rapid.

Based on NIRS, a method is being introduced that is used for measuring the N concentration in the perennial ryegrass and red fescue. The N concentration lies in the range from 0.6 to 6.26%. This concentration can be predicted with an RMSEP of 0.19 to 0.24% of N using PLSR models. These models are applied to the raw and scatter to correct the NIR spectra.

Therefore, samples from multiple years are needed to be included in the calibration data. This is done in order to increase the robustness of the models. In comparison to the reproductivity of the Dumas method, the

RMSEP is said to correspond to a higher measuring error. But the NIRS method is developing sufficiently accurate and precise in order to replace the Dumas method for evaluation of the plant N status in the field in practical seed production.

In order to have a cheap determination of N concentration in the future, the method developed can be of great value accordingly. Whereas, when novel N application strategies are based on the measurement of the plant, N concentration is developed. When the N budgets are introduced in the grass seed production.

2.4.3. Remote Sensing

Strong correlations should exist between the data from the satellite imagery and the concentration of many of the biochemicals with the vegetation canopies (Jago et al., 1999; Curran et al., 1997). The concentration of the chlorophyll within the vegetative canopy is said to be positive. This is related to the point of the maximum slope at the wavelength that lies between 690nm and 740nm in the reflectance spectra (Miller et al., 1990).

This point is known as the "red edge" of plant reflectance. The red edge characterizes the effective boundary. This boundary is between the strong absorption of red radiation that is by the chlorophyll and with the increased multiple scattering of radiation in the near -infrared wavelength (Curran et al., 1990; 1991). In order to explain the movement of the red edge to a shorter wavelength, researchers have also used the red edge chlorophyll concentration relationship. They included red edge chlorophyll as a result of senescence or stress- induced chlorosis (Schutt et al., 1984).

Therefore, a useful indicator of canopy chlorophyll concentration is provided by the information on the REP. This information can be used as an indicator of vegetation stress, development stage, photosynthetic capacity, nitrogen content, and productivity (Curran et al., 1990). Therefore, a strong correlation gets developed between the nitrogen addition and chlorophyll concentration ($r=0.82$), which is at the field site (Jago et al., 1999).

If there is an increase in the supply of nitrogen yield then there will be a shift of red edge position towards the longer wavelength. This was shown by the Mutanga and Skidmore in 2007. The red edge position, amplitude and the slope at 713 nm and slope at 725 nm were significantly correlated. This correlation measures the nitrogen concentration which includes bootstrapped $r=0.89$, -0.28, 0.63 and 0.75, respectively, even at the canopy level (Mutanga and Skidmore, 2007).

A summarized method is used to scale the biochemical information from the leaf level to the canopy level. This theory was given by He and Mui in the year 2010. According to the semi-arid heterogeneous grassland, it was concluded that all the methods are useful. But they said even none of the method is ideal.

Clevers and Kooistra in the year 2012 showed that for their Dutch study sites that the CI red edge is red edge chlorophyll index. They found that it was commonly found to be a good and liner estimator of the canopy N content. That is for both the grassland sites where R=0.77 and for the potato field with R=0.88. Their approach is basically used to apply with e.g. MERIS and Hyperion data. Their approach can also be used with the upcoming Sentinel -2 and -3 system.

2.5. IMPORTANCE OF CONSERVATION OF GRASSLAND

The strategy that is used to provide the context for the protection, management and restoration of the native grasslands in the ACT that is commonly known as the ATC Native Grassland Conservation Strategy.

The vision statement for the strategy was outlined in the previous section. According to this strategy "Healthy native grasslands supporting diverse flora and fauna for now and the future". In order to achieve this vision, the various strategies are being organized around the conservation goals that are explained in this section.

2.5.1. Protection Goals

The protection goals include various points:

1. These protection goals focus on conserving all the remaining areas of the native grassland in the ACT. These native grasslands are in a moderate to high ecological condition.
2. It works in order to retain the areas of the native grassland in the lower ecological conditions. These conditions serve as ecological buffers, landscape linkages, or even contribute significantly to the threatened species in order to conserve them or work as a priority for rehabilitation.
3. The protection goals help in conserving the viable wild populations of the native grassland. This grassland includes

flora and fauna species in the ACT and they also support local, regional, and national efforts that work toward conservation of these species.

2.5.2. Threat Management Goal

1. The threat management goal prevents or manages the impact of the threatening process. This process helps in maintaining or improving the ecological conditions and biodiversity of the native grassland. This threat management goal works with particular attention in order to threaten the species.

2.5.3. Management Goals

1. Management goals help in managing the native grassland in the ACT that works across all the tenures. This maintenance of native grassland takes place to maintain or improve the ecological condition and biodiversity. This works with particular attention towards the grassland habitat of threatened species.

2.5.4. Ecosystem Function and Connectivity Goals

1. This type of system works in native grasslands in a good ecological condition. These ecological conditions support the viable population of the grassland species. They are well connected in the landscape and are considered more resilient that includes climate changes also.

2.5.5. Monitoring and Research Goals

1. Monitoring and research goals include sound research, monitoring, and adaptive management. These support the conservation of the native grasslands and with that, support the component species.

2.5.6. Community Engagement Goals

1. Community engagement goals include community groups, landholders, and others that are actively involved in native grassland conservation.
2. For the conservation, an informed community supports and uses the native grassland areas.

2.6. CONSERVATION OF GRASSLAND

The important use of environmental science is to concentrate on the aim to conserve the grasslands. Grassland is the important ecosystem that people use. Their use is for the resource they are provided. Sometimes, the grassland or rangelands are often known as or called as an open area without any tree.

Farmers and ranchers use the grasslands in order to gaze at their animals such as cattle and sheep. When these herd animals are allowed to inhabit these open grasslands, this leads to some unexpected environmental changes and can damage these areas.

Figure 2.6. Conservation of Grassland.

Source: Image by Flickr

Overgrazing occurs when herd animals consume the vegetation that covers the grassland to the degree that it cannot get naturally replenished or recovered. Thus, after the herd animals remove the vegetation by overgrazing, these grasslands are susceptible to the destruction that happens through wind and water. This in turn is responsible for making the grassland into a desert and makes the land generally unable to regain its vegetation. This process of changing fertile land or useful vegetative land into a desert is commonly known as desertification.

There is one approach that is used to minimize ecosystem damage or desertification. This process is known as rotational grazing, which helps in overcoming the effect of overgrazing. Rotational grazing allows farmers and ranchers to herd their animals to graze in a particular area for only a shorter time and then moving them into different areas. This rotational process from one grazing spot to another is the natural process of herd animal grazing a landscape and allows the patches of the grassland to recover before the animals return to that place again for gazing.

2.7. CONSERVATION PRACTICES

2.7.1. Access Control

- **Description:** Access control limits the amount of time or time of year. This limitation is applied to vehicles and livestock that are said to have access to water bodies, considered as an environmentally sensitive area to hazardous areas.
- **Maintenance:** The barriers that are used should be periodically inspected and repair should be done as needed.

2.7.2. Brush Management

- **Description**: Brush management is used for reducing or eliminating the undesirable vegetation in order to increase the vigor, quantity, and amount of the desired vegetation that is present, and to increase wildlife habitat.
- **Maintenance**: The treatment of the spot of the individual plant. Areas should be provided re-treatment as per the requirement.

2.7.3. Fence

- **Description**: Fences are constructed barriers that are used to control animal traffic patterns. These fences are applied to reduce erosion and to control the access of grazing animals in order to permit recovery or stockpiling of vegetation.
- **Maintenance**: Fences require routine inspection, which is included as part of an on-going management program. After spring season, when the snow melts, and after storms, these fences should be inspected. This inspection is required to determine the weakness, breaks, or malfunctioning that might have affected the intended use of these fences.

2.7.4. Forage Harvest Management

- **Description**: Forage harvest management deals with the cutting and removal of forages. This process is done to optimize the yield, stand life, quality, to maintain the wildlife habitat, and to control insects and other pests.
- **Maintenance**: It is important to clear the land that might contain the debris, which may damage the machinery. This process should be done before the forage harvest. It should be noted that forages should be cut until the dew or rain on leaves gets fully evaporated. Mow is the most recent seedings that are used in the case of older stands.

2.7.5. Grade Stabilization Structure

- **Description:** Grade stabilization structure is used to control the channels' grade in natural or constructed watercourses. They may be used as a source of livestock water, control of gully erosion, and wildlife habitat.
- **Maintenance:** It is important to prepare the operation and maintenance plan for the person that is responsible for operating and maintaining the system. There should be a proper plan for periodic inspections and speedy repair or replacement of the damaged components.

2.7.6. Heavy Use Protection Area

- **Description**: The heavy use protection area helps in stabilizing the areas that are frequented by vehicles or livestock. This is done in order to reduce the erosion in or near the critical water bodies, improve the herd health, and to improve the water quality.
- **Maintenance**: The maintenance takes place by an operation and maintenance plan. These plans are prepared and reviewed by the landowners or operators.

2.7.7. Nutrient Management

- **Description:** In order to increase the forage production, proper placement of the correct amount of nutrients at the correct stage of the plant growth is necessary. It is also important to reduce the amount of loss nutrient to the surface or groundwater sources.

These are also used to increase production and gain more profits.

- **Maintenance:** Equipment needs to be calibrated to ensure the uniform distribution of the material at the planned rates. There is the requirement of documents to apply the actual rate of the nutrients. Changes in the animal number or feed management will necessitate the additional analysis.

2.7.8. Prescribed Grazing

- **Description**: Prescribed grazing helps in managing the harvest of the vegetation. This harvest management takes place in order to maintain or improve the desired plant community. This process is also used to maintain the groundwater quality, reduce erosion, and that to improve the cover for wildlife.

Figure 2.7. The image representing Grazing methods of livestock.

Source: Image by Deposit photos

- **Maintenance**: In order to ensure the objectives are actually met, there is a requirement of monitoring the data and grazing records on a regular basis. Necessary changes should be done in order to prescribe the grazing plan to meet the objectives.

2.7.9. Riparian Forest Buffer

- **Description**: A riparian forest buffer is an area that is used to establish trees or shrubs which are adjacent to a stream, lake, or other water bodies. This process is done to improve water quality, reduce sediment delivery, create shade for the aquatic habitats, mitigate the flood damages, etc.

- **Maintenance:** In this method, the trees that come under the buffer area need to be periodically maintained and harvested. Tree harvesting is important for the plant health and buffer function as the buffer matures.

2.8. CONSERVATION ACCORDING TO THE REGIONS

The approach is to focus on the regional conservation planning for the temperate grasslands. The planning will differ from region to region. These variations are determined by factors like the current level of protection and disturbance, the existence of the ecological condition and potential that is required to have the connectivity, the possibility of restoration, the status of the wildlife population, and ability for reintroduction. There some other important factors like prospects for sustainable management and further protection. These factors are used to lead to the quality of the template grasslands that are classified as large and relatively intact landscapes. These intact landscapes arc highly modified and fragmented landscapes remnants.

It is biologically important that with large landscape conservation, attention is also given to these temperate grasslands that are modified and fragmented. The maximum amount of retention of biological diversity should be ensured on these grasslands. With this, it also important to ensure the long term potential that is used for connectivity and the ability for indigenous people to retain their culture and livelihood.

The urgency for legal protection cannot be overstated in grasslands that are being highly modified. In these grasslands, it is said that less than 5% of the original ecosystem is remaining. This is the case in the steppes of eastern Europe, tall grasses prairie of North America, and the low land grasslands that are present New Zealand and south-eastern America. There is recent literature that reviews the total economic value of the indigenous temperate grasslands. This literature, sponsored by TGCI, found no studies that will specifically address the empirical value of these grasslands. It is assumed, traditionally, that when the grasslands are treated as a commodity and are sold on the market place. Temperate grasslands are in all likelihood. The latest understanding of the biome explains that there is total value to sustainable economic use, their contribution to human health and well-being, and ecosystem goods and services. If the economy is not fully able to provide an accurate picture of the net benefits of the temperate grasslands, poor land use and investment decision, then that county will result, and their important value of the grassland is said to lose to the society.

2.9. CONSERVATION SIGNIFICANCE CATEGORIES

2.9.1. Floristic Value Score

In order to protect and manage the remaining native grasslands in the ACT, each of these known sites is being assessed and categorized into the Conservation Significance Category (CSC). In the previous strategies, the ACT government has been adopted. In this, they use the Floristic Values Score (FVS). FVS is used to quantify the native grassland ecology condition (Rehwinkel 2015; Commonwealth of Australia 2016a).

Grasslands that are in a higher condition have higher native plant diversity and are considered to have a lower level of disturbance or modification with fewer weeds present. This relativity is recently developed which measures the widely used assessing grassland condition.

The definitions of the major grassland classification include 'Grassland', 'Native Grassland' and 'Natural Temperate Grassland'

A grassland is considered as a part of the natural temperate grassland of the southeastern highlands. They can be considered as a critically endangered ecological community under the EPBC Act (Rehwinkel 2015; Commonwealth of Australia 2016a) if:

- it is dominated by the native vascular plants
- it meets the requirement of non-grass and indicator species composition
- it has an FVS of 5 or more

The FVS plays an important role in allowing an area of native grassland to be assigned to an appropriate CSC. The measures that determine the CSC of the patch is equally important with the ecological conditions of the grassland. These patches are of size and value as habitat for the threatened species.

It should be noted that the maximum patch size is considered to contain the threatened Natural Temperate Grassland ecology community. This grassland ecology community works under the EPBC Act that contains 0.1 hectares of area.

Grasslands that do not meet the EPBC Act criteria for the natural Temperate Grassland are classed as a native grassland. This classification is given when the grassland is dominated (>50% cover) by perennial native species.

2.9.2. Key Threatened Species Habitat

This strategy involves an area of habitat that is expected to be crucial for the conservation of threatened species in the ACT. This is considered as a 'key' habitat.

If there is an equal loss of key habitat then it likely to jeopardize the long-term conservation for the threatened species in the ACT. Key habitats are usually considered to be a large area that contains the largest population of the threatened species in the ACT. They should be in a good ecological condition and contain the proper habitat for multiple threatened species.

2.10. SPECIFIC THREATS TO GRASSLANDS

2.10.1. Poor Agriculture Practice

Poor agricultural practices can ruin the soil and strip grasslands of life. The rotation of the crops is equally important for maintaining the fertility of the soil.

The rotation of crops prevents the soil from becoming infertile and preventing anything from growing in that area for several years.

2.10.2. Monocropping

Monocropping is growing only one crop at a time. This is an agriculture practice which depletes soil nutrients. Grassland thrives off biodiversity of plants and animals, but because monocropping provides only a single type of plant that tends to weaken the biome. Monocropping also increases vulnerability to natural disasters.

2.10.3. Toxic Pesticides

Toxic pesticides are used in agricultural crop land. This can play a deadly role for wild flora and fauna.

2.10.4. Grazing Livestock

Grazing Livestock can consume, trample, and destroy grasses. Selective grazing can reduce the competitive nature of the entire ecosystem. This grazing method effects by weeding out some of the plants and allowing others to overpopulate.

Figure 2.8. Grazing animal for food in a free vegetative grassland.

Source: Image by Farmers Weekly

In addition to this, conventional agricultural cropland provides few food sources and nesting areas for birds.

2.10.5. Continued Global Warming

Climate change could turn the current marginal grassland into desert. This is because the rainfall pattern changes.

2.10.6. Development of Urban Areas

The development of urban areas leads to the reduction of grassland habitat. This effects by decreasing the amount of vegetative grassland.

2.10.7. Invasive Species

There are some invasive species that can displace native plants and reduce the quality of the grassland. The invasive plants may not be equipped to handle the extreme weather which includes droughts, wildfires, thus they result in habitat loss.

2.11. SOLUTIONS TO GRASSLAND PROBLEMS

- There should be continued efforts towards education. The education should be particularly provided to farmers, and should focus on protecting the soil and preventing soil erosion.
- It is important to protect and restore the wetlands. Protection and restoration are important parts of grassland ecology.
- Rotation of agriculture crops is equally important. This helps in preventing the sapping of the nutrients.
- Planting of trees is also very important because it helps in windbreaking, which reduces erosion on farm fields. However, it is crucial to choose the right plant species according to the area.
- Controlling the conduction of dry season burning as this stimulates the fresh plant growth and restores the calcium to the soil. This helps in building up in the dry grasses. Some species of native plants need the occasional fire to thrive. The blazes can also help in removing the invasive species.

2.12. CONCLUSION

There are certain methods that are used for the production of grasslands. These processes include the usage of efficiency of radiation that includes the production of plants using chlorophyll. Similarly, there are many other ways that are used for the production of grassland.

There are different methods that are used to measure the production of a grassland. These procedures are measuring the unit of yield, destructive measurement with the non-destructive measurements that tell how destruction on non-destruction can help in maintaining the grasslands. There are some remote sensing procedures that involve monitoring of land that includes classes.

It is important to maintain the nutrient of the grassland. As the production of the grassland is important for society, similarly it is important to maintain those production processes and maintain the grassland.

There are different countries which contain different areas. These areas require specific growth parameters for maintaining the grassland. During the production of grassland or maintenece of the grassland, it is important to first calculate the factor that is responsible for maintaining the grassland in that particular area.

There are some specific threats of grasslands that effect the growth of the vegetative grassland. Those threats need to be addressed so as to have good vegetative growth. These factors include poor agricultural practice, monocropping, use of toxic pesticides, grazing livestock, development of urban areas, invasive species, and many other factors.

REFERENCES

1. ACT NATIVE GRASSLAND CONSERVATION STRATEGY AND ACTION PLANS. (2017). [ebook] Available at: https://www.environment.act.gov.au/__data/assets/pdf_file/0010/1156951/Grassland-Strategy-Final-WebAccess.pdf [Accessed 19 Feb. 2020].
2. Grassland Conservation and Management. (n.d.). [ebook] Illinois. Available at: https://www.inhs.illinois.edu/files/3113/9483/0974/GrasslandSciencePolicy.pdf [Accessed 19 Feb. 2020].
3. Grassland Conservation Practices for a Sustainable System. (n.d.). [ebook] Available at: https://www.nrcs.usda.gov/Internet/FSE_DOCUMENTS/nrcs142p2_006488.pdf [Accessed 19 Feb. 2020].
4. Henwood, B. (n.d.). *Towards a Conservation Strategy for the World's Temperate Grasslands*. [ebook] Available at: https://www.iucn.org/sites/dev/files/import/downloads/tgci_strategy_paper_2010.pdf [Accessed 19 Feb. 2020].
5. National Geographic. (n.d.). *Grassland threats, explained*. [online] Available at: https://www.nationalgeographic.com/environment/habitats/grassland-threats/ [Accessed 19 Feb. 2020].
6. NATURE. (1969). [ebook] Available at: https://www.nature.com/articles/221691a0.pdf [Accessed 19 Feb. 2020].
7. Schils, R., Velthof, G., Mucher, S., Hazeu, G., Oenema, O., Wit, A. and Smit, A. (2013). *Methods to estimate grassland production and biological fixation*. [ebook] Wageningen. Available at: https://ec.europa.eu/eurostat/documents/2393397/8259002/Grassland_2014_Task+2.pdf/42873c7f-dfdf-49ca-b2ef-7c7b5bcabfc8 [Accessed 19 Feb. 2020].
8. Curran et al., 1990
9. Curran et al., 1990; 1991
10. He and Mui (2010
11. Jago et al., 1999
12. Curran et al., 1997
13. Kara, 1998
14. Larcher, 2003; Jing et al., 2011
15. Miller et al., 1990
16. Mutanga and Skidmore (2007)
17. Rehwinkel 2015

18. Schutt et al., 1984
19. Simon and Lemaire, 1987
20. Sinclair and Weiss, 2010
21. Smit et al., 2008
22. Temminghoff et al, 2000

3

Impact of Climate Change on Grassland

CONTENTS

The chapter of impacts of the climate on grassland explains the basic significance of the impact of the climate change on grassland production and soil carbon all across the globe, which includes grasslands and the global carbon cycle, and evaluation of the grassland response to climate change. This chapter also explains the impacts of the increased atmospheric carbon dioxide concentration and carbon monoxide climatic interaction such as detecting changes in the grasslands.

This chapter also explains the effects of climate change on grassland biodiversity and productivity which emphasize a summary of model types. This chapter addresses the impact of the climate change on the grassland such as impact of the climate change, impact of the climate change and atmospheric carbon dioxide concentration on species composition, and the impact of climate change and atmospheric carbon dioxide concentration on grassland nutrition.

This chapter also mentions the shifting plant species composition in response to climate change stabilizes grassland primary production. This chapter also includes the climate change effects on forests and grasslands such as benefits/stresses, effects, water/precipitation, water/snow cover, water/flow, water/sea-level rise, and vegetation/growing season.

3.1. INTRODUCTION

Sudden changes in the climate impact with respect to the grasslands and prairie bioregions is consisting of increment in the seasonal, annual, minimum and maximum temperature and altering perception patterns. Due to these ecosystems are comparatively dry with a strong seasonal climate, they are very sensitive with respect to the climatic alterations and vulnerable as well in the context of changes in the climate regime.

For instance, model simulations depict that the regional drought in the Prairie Pothole Region may result in the loss of precious habitat for breeding the water flow, from a region that historically has harvested near about fifty percent to eighty percent of ducks of the continent. Hotter as well as dryer circumstances may also take into the direction towards the violation of the new species, and a higher chance of risk of wildfire.

Increment in the temperatures, lesser rainfall, and drought are already being noticed in several numbers of regions, and the arid ,especially in Southwest, is projected in order to become even drier in this century. The forest region that is wetter is likely to intrude the existing savannas, on the other hand, increment in the arid region deserts projected in order to expand to some level and move upward in elevation, as an outcome this produces "desertification" of the arid grassland ecosystems.

One of the most major aspects that impacts the grassland and prairie ecosystems is Drought. Drought is likely to be led to the increment in the wildfire as well as the loss of wetland habitats, for example, prairie potholes that are very important habitat with respect to the migratory bird species, along with the species migration and habitat shifts.

Figure 3.1. The basic significance of the climate change on grassland.

Source: Image by PxHere

Vegetations shifts from C3 to C4, the communities of grassland and phenological changes will influence or impact the ecosystem as well as species, and the modification in the composition of species along with the productivity of the plant that may also influence or affect the human communities that depend on the agricultural production in these kinds of areas. According to the NASA satellite imagery shows that the anticipated modifications in plant productivity are already taking place all across the globe.

A slight modification in the temperature and rainfall can change the composition, abundance, and distribution of species in arid lands in a considerable amount, and the products as well as services they offer.

For instance, observed and projected decrement in the regularity of freezing temperatures, lengthening of the frost-free season, and increment in the minimum temperatures can change the species of plants and the alteration of the geographic and elevational boundaries of several numbers of arid lands.

The level of these alterations or modifications will also rely on the shifts in precipitation and fire. Increment in the frequency of the draught could also produce major changes with respect to the vegetation. Losses of vegetation cover that are integrated with the increases in the intensity of the precipitation and climate-induced reductions in the soil aggregate the stability will enormously increase the rates of erosion.

Transport of eroded sediment to streams integrated with the alterations in the timing and the amplitude of the minimum as well as maximum flows can impact the quality of water, aquatic fauna, and riparian vegetation. Land trusts and the conservation agencies are already observing these kinds of shifts and are working in order to identify the risks, decrement in vulnerabilities, and regulate the grassland systems for changing conditions.

3.2. IMPACT OF CLIMATE CHANGE ON GRASSLAND PRODUCTION AND SOIL CARBON WORLDWIDE

The effect of the sudden changes in the climate and the increment in the atmospheric $CO_2$was modeled for 31 temperate and tropical grassland sites, with the help of the CENTURY Model. Sudden changes in climate elevated the net primary production, except in cold desert steppe regions, and CO_2 increased production everywhere. Sudden changes in the climate produce soil carbon to decrease overall, with a loss of 4Pg from global grasslands after the time of five decades.

Both sudden changes in climate and increased CO_2, altogether, results in increment in the production and decrement in the global grassland carbon losses to 2Pg, with tropical, savannas becoming small sinks for soil C.

Determination of statistically significant modification in the production of plants would require a change of near about 16 percent, which is measured in the production of a plant because of high year to year variability in the plant production. Most of the anticipated alterations in plant production are less than ten percent.

3.2.1. Grasslands and the Global Carbon Cycle

The potential impacts of the sudden changes in climate on the grassland biochemistry have received much less attention as compared with the forests (Hall & Scurlock 1991). The impacts of the modifications or alterations in the temperature, water as well as nutrients are comparatively well acknowledged, and the interactions with the long-term impacts of the carbon dioxide fertilization are now starting to be sufficiently well-developed in order to calculate the alterations in the carbon fluxes (Schimel et. Al. 1990, 1991; Long 1991; Owensby et. Al. 1993).

Figure 3.2. Grasslands and the global carbon cycle.

Source: Image by PxHere

Grassland ecosystems store most of their carbon (C) inside the soils, where turnover times are comparatively long (hundred years to ten thousand years), and so alterations, never the less, they can take place slowly, will be of the considerable interval of time. Alterations in the grassland carbon storage will have a significant as well as long-lived impacts on the carbon cycles all across the globe.

Fluxes of carbon with the help of plant and soil organic matter in grasslands, more particularly in the tropical regions, may have been underestimated in the plants (Long et. al. 1989). Grasslands are one of the most extensive types of vegetation all across the globe, which covers near about twenty percent of the land surface of the world's land, that is (24 x 106 km2).

Grasslands are consisting of more than ten percent of the soil carbon stocks all across the globe (Eswaran et al.1993). Grasslands are also a significant or meaningful habitat with respect to the wildlife, humans ,as well as their domestic livestock, and are subject to conversion to both irrigated agriculture as well as dryland.

Under the guidance of the integrated SCOPE (Scientific Committee on Problems of the Environment) project (Breymeyer & Melillo 1991), the CENTURY model of the interactions of the soil and plant has been utilized to an extensive range of grassland all across the globe (Parton et at. 1993).

The SCOPE- GRAM group has accumulated or collected long term data from near about thirty-one sites in the context of both tropical grassland as well as temperature. These collected data are then used to evaluate the robustness of the CENTURY model and the simulated response of the grasslands with respect to the scenes that are related to the sudden changes in the climate.

Various numbers of recent studies have approximated the response of the soil organic matter (SOM) in the context of environmental change. Jenkinson et al. (1991) approximated the considerable as well as quick emission of carbon from soils, which, as an outcome produces global warming (61 Pg over the time span of six decades).

On the other hand, Schlesinger (1990) debated that the undisturbed soils will have a low potential for added carbon storage and can only change with a slow rate.

According to most of the previous studies, they have had recognized the limitations or disadvantages. Among from that, studies done by the named researcher Schlesinger, who used inferential soil chrono-sequence studies.

On the other hand, the model of Jenkinson et. al. (1991) did not well think out the feedback between the soil processes and the sudden changes in the climate impacts the primary production, which is a prime aspect of CENTURY.

Certainly, Jenkinson et, al. (1991) observed that, in order to find out how the sudden changes in climate are impacting the flux of carbon dioxide (CO_2) between the soil and the atmosphere, their model 'would have to be integrated with the another which specifies how plant production all across the globe is changed with the help of increment in the concentrations of the greenhouse gases (Jenkinson et al. 1991).

3.2.2. Evaluation of Grassland Response to Climate Change

Sites were selected in order to represent an extensive range of the regions of the grassland, which is based on the climate factors as well as properties of the ecosystem. This is consisting of near about fifteen temperatures and the tropical regions for which detailed data on a monthly basis were available on the vegetation biomass and the plant debris.

A detailed comparison of the simulated and observed peak live biomass as well as the plant production for these fifteen regions, depicts a good correlation (N = 0.45 and 0.70, respectively).

Soil carbon level was anticipated very well (N=0.93). Researchers has assembled at least twenty-five years of weather data on a monthly basis for fifteen various sites and a further sixteen grassland sites that are derived from the literature. The sites are ranged from temperature ecosystems with patterns of highly seasonal temperature to the tropical areas where the temperature fluctuates very little.

Precipitation at all of these regions showed a marked seasonality, and in most of the scenarios, a high level of variability has been shown in the time period of several years.

With respect to the temperature, seasonality of the climates, and rainfall, there are a total of thirty-one sites into seven bioclimatic areas that are based on a modification of Bailey eco-regions, that are well thought out as ecological factors as well as climatic factors (Bailey 1989) (as shown in the Tables 3.1. and 3.2. mentioned below).

The three divisions of Bailey's dry domain are based on the temperatures, with the cold desert steppes that are having a regional mean annual temperature of near about -0.3 degrees celsius. The temperate steppes having near around 10.2 degrees celsius and the dry savannas are near around 24.1 degrees celsius.

Table 3.1. Long-term mean annual precipitation, temperature, potential evapotranspiration rate, and the ratio of annual precipitation to potential evapotranspiration rate (PPT/PET), for the seven grassland regions based on Bailey (1989)

Biome type	Mean annual precipitation (mm)	Mean annual temperature (°C)	Mean annual potential evapotranspiration	PPT/ PET	Global area (1o6 \ 2)	Vegetation type
Cold Desert Steppe	299	-0.3	960	0.334	2.095	C3
Temperate Steppe	301	10.2	1041	0.294	2.943	C3 / C4
Humid Temperate	700	10.2	1041	0.294	2.943	C4
Mediterranean	497	15.7	1253	0.401	0.161	C3 C4

Dry Savanna	387	24.1	1977	0.165	5.106	C4
Savanna	788	23.4	1491	0.554	7.990	C4
Humid Savanna	1555	27.4	1404	1.195	1.708	C4

Source: Table from Reseacrchgate.net

Table 3.2. Simulated mean annual above- and belowground plant production, total SOC, mean annual abiotic decomposition factors, and net N mineralization rates for each of the seven grassland regions prior to climatic changes

Biome type	Aboveground production (gm $^{-2}$)	Belowground production (8*) ^{2}	Total Soil C (0-20 cm) (gm $^{-2}$)	Net N mineralization (g N m* 2y*')	Abiotic decomposition rate (0-1)
Cold Desert Steppe	60	92	5530	3.11	0.093
Temperate Steppe	36	59	2020	2.67	0.158
Humid Temperate	134	161	7171	4.81	0.194
Mediterranean	79	103	3005	3.44	0.184
Dry Savanna	55	85	2275	2.58	0.173
Savanna	191	222	4306	5.73	0.339
Humid Savanna	340	344	3273	6.61	0.644

Source: Table from Researchgate.net

Model Runs Under Climate Change Conditions

The doubled carbon climatology that is utilized in order to drive the CENTURY model in the present study were derived from two Global Climate Models (GCMs), the Canadian Climate Centre (CCC) and the Geophysical Dynamics Laboratory High Scenario (GFHI). These models are used to offer assistance as they provide monthly projections of the temperature and the precipitation for 2.5 degrees grid cells all across the globe.

The approximated values for a doubling at atmospheric carbon concentration, and assumed that these would be reached within the time interval of near about five decades from the present interval of time, corresponding roughly to IPCC Scenario A (Houghton et al. 1990). Anticipating alterations with respect to the annual as well as seasonal temperature and precipitation.

The two global climate models simulate identical alterations with respect to the annual air temperature (2-5 degree Celsius). Never the less, the alterations in the precipitation were different between the models. The Geophysical Dynamics Laboratory High Scenario (GFHI) model depicted the increment in the precipitation for all of the areas. On the other hand, the Canadian Climate Centre (CCC) model shows the decrement in the precipitation for the temperature steppe and humid savanna regions and the increment for the other regions.

3.3. EFFECT OF INCREASED ATMOSPHERIC C_2 CONCENTRATION AND CO CLIMATE INTERACTION

The effects of the increment in the atmospheric carbon concentration on the grassland ecosystems with and without the inclusion of the sudden changes in the climate. Overall, the impact of the increment in the atmospheric carbon concentration under the present patterns of conditions of climate was to increase the total plant production, the rate of abiotic decomposition and the SOC storage. On the other hand, nitrogen mineralization rated falls.

Production elevated the most of in the cold desert steppe region, and SOC storage elevated most for the humid savannas. Soil carbon increases were negatively correlated with the alterations in the rates of the decomposition.

Nitrogen mineralization rates are fallen in several numbers of regions, as an outcome from an elevated input of the lower quality litter with the higher carbon to nitrogen ratio (Melillo et al. 1982), but elevated for the cold desert steppe region.

The stimulated elevates in the decomposition under the increased carbon, resulted from the enhanced soil water relationships at the cold dry sites (cold desert steppe, temperature steppe), with the smaller increment for the warmer and more humid regions. When the relative alterations in the plant production, SOC, nitrogen mineralization, and the decomposition were compared between

1. The integrated effects of the elevated atmospheric carbon (C_2) concentration and climate and
2. The independent effects or influences of carbon C_2 and climate change, a linear additive effect was demonstrated (N = 0.97, 0.98, 0.97 and 0.91, respectively).

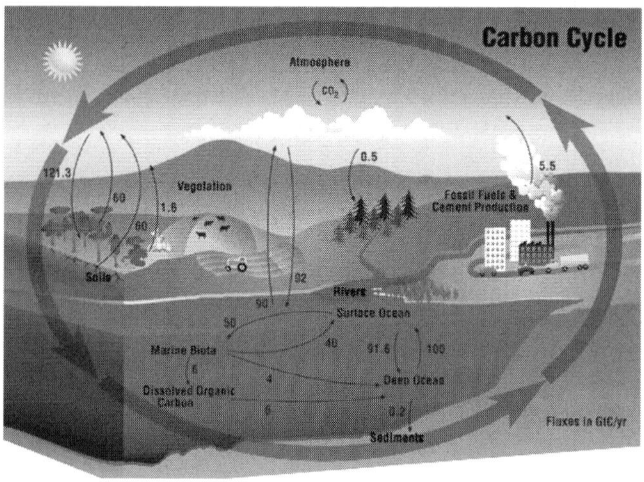

Figure 3.3. Effect of increased atmospheric C_2 concentration and CO climate interaction.

Source: Image by Flickr

Elevated atmospheric carbon C_2 concentration helps in improving the total plant production regardless of the impacts of the sudden changes in climate, consistent with the results of Melillo et al. (1993).

This improvement in plant production helps in reducing the SOC losses which as an outcome from climatically driven alterations in SOM decomposition. The direct carbon C_2 impacts the decrement in carbon losses all across the grassland systems and results in a SOC sink region in the tropical savanna and humid savanna areas, irrespective of climatic impacts.

The effects of this climate and carbon alter the present grassland SOC stocks show that the climatic effects alone will produce a net loss of near about 4 Pg carbon (as shown in table 3.3. mentioned below). With all regions becoming carbon sources as an outcome of decomposition of SOC. Never the less, the effect of the increment in the atmospheric carbon $C_2$is to offset the SOC losses due to the sudden changes in climate.

The effect of carbon C_2decreases the net losses all across the globe from grasslands by near about fifty percent in the Canadian Climate Centre case and sixty-six percent in the Geophysical Dynamics Laboratory High Scenario case, recommending a net carbon source of 2 Pg C, over the time period of five decades, or 0.04 Pg C per annum.

Talking about the tropical regions, savannas along with the humid savannas actually become SOC sinks with the inclusion of Carbon $C_2$and the sudden changes in the climate altogether, irrespective of the condition of GCM.

These outcomes fluctuate with those of Jenkison et. al (1991), who approximated a net source of 61 Pg C, all across the globe, over the time period of next six decades from the soil organic matter, all across the world.

Grassland soils (such as temperate steppes and tropical savannas) that accounted for near about twenty percent of the world SOC stocks in this analysis, recommending a grassland source of near about 12 Pg C, over the time period of six decades, or 0.2 Pg C per annum, that is based on calculations of Jenkinson.

3.3.1. Detecting Changes in Grasslands

The potential to determine the effect of the sudden changes in climate with respect to the production of plant and SOC levels, given the observed natural variability, was assessed with the help of CENTURY model results.

The twenty-five average plant production and the levels of SOC for all of the thirty-one sites that are prior to the sudden changes in climate were compared with the average values for the period of twenty five years following the start of the changes in climate for both of the GCM situations. Along with that, a statistical f-test was used in order to find out, if the mean values were considerably different at the five percent level.

Near about eight percent of the 62 model runs showed the considerably different plant production (predominantly those with the low productivity), with the alterations in the plant production averaging +22 percent (that ranges from seventeen percent to near about thirty-one percent). The non-significant comparisons had an average shift in the plant production of +7 percent (that ranges from zero percent to sixteen percent).

Around 81 per cent of the comparisons depicts the considerable shifts in the mean SOC, with SOC changes averaging seven per cent (that ranges from one per cent to near about -33 per cent). SOC changes of less than one per cent were non-significant.

These outcomes recommend that the stimulated SOC and the plant production would have to shift by the value of at least one percent and sixteen percent, in order to be statistically detectable. Noting that it would be impossible to determine one percent change in SOC in the field because of the large spatial variability in SOC (Yonker et. Al 1988) and it would take near about five percent to fifteen percent to be detectable.

According to the several numbers of comparison that shows a considerable amount of fluctuations or variations with respect to the plant production and SOC did not change with time (for the time span of 25-50, 50-75, and 75-100, following the start of the climate change).

Modifications in the plant production or SOC that is induced with the help of sudden changes in climate and increment in the content of carbon present in the atmosphere will, therefore, have to be considerable, overcoming inter-annual variability and sampling variance, respectively, in order to be detectable.

A comparison of the anticipated alterations in the plant production with this statistical analysis recommends that most of the anticipated shifts in the plant production would be very hard to find out. Never the less, the plant production has stabilized seventy-five years after the start of the climate change, SOM has not reached a new steady- state, even though the rate of alteration in the SOM has been reduced.

Comparison of the transient and long term (near about 5000 years) equilibrium levels of SOM depicts that only twenty-six percent of the long-term shifts in SOM has taken place near about fifty years to seventy-five years after the initiation of the alterations in climate. In this way, it is very important to differentiate among the transient responses that are induced instantly with the help of sudden changes in climate and the long-term equilibrium response.

According to Bailey's humid tropical domain that was divided into the savannas and humid savannas on the basis of precipitation, the former receiving a regional annual mean of near about eight hundred millimeters and after that, it becomes near about 1550 millimeter. The simulated outcomes for each bioclimatic region were assessed with the help of averaging results from the regions that are in a particular bioclimatic zone.

3.4. EFFECTS OF CLIMATE CHANGE ON GRASSLAND BIODIVERSITY AND PRODUCTIVITY

There is an increment in the proof that the influence of the sudden changes in the climate with respect to the productivity of the grasslands will at least partly relies on their biodiversity. A high level of biodiversity can discuss the stability of the grassland ecosystems against the environmental change, but there are also direct impacts of the biodiversity on the quantity as well as the quality of the productivity of the grassland.

Figure 3.4. Effects of climate change on grassland biodiversity and productivity.

Source: Image by Pixabay

In order to explain the manifold, the interactions, and in order to anticipate the future climatic responses, models can be used. Never the less, model designed for the study purpose, the interaction between biodiversity and productivity tends to be structurally different from the models for studying the effects of the climatic effects.

Permanent grasslands are every so often hot spot of biodiversity, which takes participation in the temporal stability of their services. The variety of the plant species that are present in the grasslands is strongly impacted by the long-term management practices, with livestock grazing being the major driving force that is impacting the vegetation dynamics, species distribution and the landscape-scale biodiversity in addition to forage quantity and quality.

Biodiversity, which is surrounded by several types of species and across landscapes, can be crucial for the longer-term resilience of the ecosystem functions and the services that they underpin. Biodiversity is closely connected to the ecosystem services with the help of several numbers of different relationships, with the communities that are full of species are tending to perform better as compared to any other individual species.

Biodiversity-ecosystem functioning relationships are impacted with the help of several numbers and identities of the species, their evenness within the community, their functional characteristics, and their interactions. If the species loss can be compensated with the help of other species that are taking participation similarly to functioning, the role of the individual species may shift with the environmental change.

Then, abiotic alterations can lead to the biotic alterations and vice versa, and the various kinds of species that take participants to most to any given ecosystem process at different points in the time and space. Never the less, there are limits to the species redundancy, and high biodiversity is required in order to maintain the several kinds of processes that are operating in multi-functioning ecosystems.

As concepts of the functioning of the ecosystem have evolved, work has broadened to surround the biodiversity loss withing and between the trophic levels. This work has offered several advantages from the studies of food webs and more extensively on ecological networks.

Never the less, according to the researcher named as Novak, who concluded that these kinds of studies can only offer limited predictive capacity. On the other hand, the general knowledge of the strengths of interactions between several kinds of species remains poor.

The requirement for the elevated predictive capacity is pressing, as the ecosystem of the world undergoes unprecedented modifications with the species being lost from an extensive range of the ecosystems and the trophic levels and there is a requirement in order to well thought out the role that models can play.

Particularly speaking, the challenges that grassland systems are coping up in the present interval of time, which shows that the aspects that are related to the diversity of the plant that cannot be ignored in modeling studies. In this way, grassland models are required to consider the nutritional value of the multi-species swards.

The differing herbage intake of grazing animals between mixtures and the pure stands, the dependence of the content of the milk protein on the botanical composition of swards, the beneficial effect of legumes on the nitrogen economy of the multi-species swards, and the supporting and controlling services (for example, pollination, drought resistance, and pest control) provided with the help of multi-species swards.

With respect to the primary importance, the requirement in order to account for the complicated and variable relationship between the grassland biodiversity and productivity. If primary productivity is not used with the help of grazers or humans, then it may sustain the larger biodiversity, never the less, some level of grazing may stimulate biodiversity.

In addition to this, decreasing the intensity of the management (for example, fertilization) in grasslands decreases livestock productivity but may improve the quality of the meat. In this way, trade-offs between the diversity and the way grassland productivity are exploited exist, and sudden changes in the climate can impact how to prioritize on or the other. Modeling tools are required in order to negotiate these trade-offs.

The literature on grassland modeling is extensive. Google scholars find more than 2 million articles that are consisting of the terms such as 'model' as well as 'grass' or 'grassland', and 1670 of those have the terms in their title. Web of Science finds 1403, in this way, titles papers (information retrieved on 20 November 2017).

The primary interest here is in modeling the responses of grasslands of different kinds of levels of biodiversity with respect to the sudden changes in the climate. Of course, the climate not only impacts the grassland biodiversity and productivity but is itself impacted with the help of dynamics at large spatial scales.Grasslands, such as all vegetation, impacts the climate with the help of their albedo and greenhouse gas balance and also can play a vital role in mitigating the climate change. The impacts of the vegetation on the atmosphere are depicted in the latest generation of GCMs (Global Climate Models), but are not further discussed in this chapter.

Proper consideration is required for both static as well as dynamic modeling, and modeling is objected at explaining the observations along with the modeling is objected at predicting the influence of environmental change. After a brief review of empirical models, process-based models, and integrated grassland modeling methodologies. These are reviewed separately, before discussing their relative strengths and weaknesses, and the scope for using elements from one model type in another. A summary of this model overview is provided in Table 3.3. as mentioned below.

Table 3.3. Summary of model types

	Empirical Models	Process-Based Models			Integrated Models
		Ecological	**Biogeochemical**	**Agricultural**	
Characteristics	Statistical models (regression etc.)	Small models focusing on biotic interactions; elegant mathematics	Parameter-rich models of cycling of carbon, nutrients, water	Parameter-rich models of phenology and yield formation	Hierarchical models that include process-based models and human agents
Static/Dynamic	Static	Dynamic	Dynamic	Dynamic	Dynamic
Strengths	Flexibility; ease of use; spatial heterogeneity	Explaining general ecological patterns including species interactions	Long-term prediction of biogeochemical cycles and soil pools	Short-term prediction of productivity	Comprehensive; scope for assessing policy
Weaknesses	Non-predictive outside calibration range	Abiotic effects and spatial heterogeneity poorly accounted for	Limited capacity for yield forecasting; no spatial heterogeneity/biodiversity	No long-term predictive capacity; no spatial heterogeneity/biodiversity	Weaknesses inherited from component models; difficult to parameterize and test

Source: Table from MDPI

3.5. IMPACT OF CLIMATE CHANGE ON GRASSLAND

Grassland accounts for 70% of the UK agricultural land and approximately 32% of agricultural output (IGER, 2003). Increased concentrations of atmospheric CO2 and changes in climate in the future are likely to have an impact on grassland and the production of grazing animals. Effects might include changes in the grazing season, yield, species and nutrient composition.

3.5.1. Impact of Climate Change

The UKCP09 future weather projections suggest overall warming accompanied by wetter winters and dryer summers (Jenkins et al., 2009). Grassland is sensitive to drought (MAFF, 2000) and dryer summers in the future might place limitations on grass growth on soils with low soil available water in the summer months (St. Claire et al., 2009).

More drought -resistant species might be required to meet livestock grazing needs or increased costs associated with additional feeding might occur. Drought will also increase the risk of wildfires affecting extensive grazing areas.

Figure 3.5. Impact of climate change.

Source: Image by Flickr

Wetter winters may cause water -logging and problems of poaching, possibly resulting in the animal having to be housed, and reduced opportunities for grazing and harvesting (Hopkins and Del Prado, 2007). Good soil structure to promote resilience to the effects of drought and water-logging should be encouraged (IGER, 2004).

Draining water- logged soil would encourage early grass growth, but at the expense of a reduction in water availability in July and August (Armstrong and Castle, 1992).

Temperature rise is likely to increase the length of the grazing season (Topp and Doyle, 1996a; IGER, 2003) with grass production starting earlier in the spring and continuing later into the autumn. This could have advantages for the availability of grass during the early lambing.

Although the rate of photorespiration increases with temperature (Nowak et al., 2004), the potential for increased herbage will be limited by nitrogen availability and in some areas irrigation during dry months and drainage during wet months (IGER, 2004).

The longer growing season may result in increased grassland yield (ADAS, 1997) though other models suggest that the longer growing season might only increase clover production and have little effect on yield overall after the impact of the drought is taken into account (Topp and Doyle, 1996a; Grime et al., 2008).

3.5.2. Impact of Climate Change and Atmospheric CO_2 Concentrations on Species Composition

Alterations in the temperature, alteration in rainfall, and atmospheric concentration of carbon dioxide (CO_2) might impact the different forage species in several numbers of different ways, in this way, changing the composition of the grassland community. Overall conditions are likely to become more favorable for forage legumes (Soussana and Luscher, 2007).

Management factors like frequency of defoliation will also be important in finding out that how the species diverse communities respond to alteration in climate and atmospheric carbon dioxide concentration (Teyssonneyre et al., 2002; Harmens et al., 2004).C3- Grasses, that have a higher content of protein as compared to the C4 grasses and in this way, they have higher nutritional value, and will help in order to elevate the competitive advantage over the C4 grasses under the conditions of elevated atmospheric carbon dioxide concentration (Ehleringer et al., 2002; Soussana and Luscher, 2007).

Never the less, the anticipated increment in the temperature will favor the C4 grassland species (Soussana and Luscher, 2007), in this way, the outcome remains difficult in order to anticipate. Generally speaking, the grasslands that are rich in species are likely to be more productive with respect to the increased levels of the atmospheric carbon dioxide concentration as compared to the species-poor grasslands (Reich et al., 2001).

3.5.3. Impact of Climate Change and Atmospheric CO_2 Concentrations on Grassland Nutrition

The prime factors that finds out the nutritional quality of grassland are; dry matter content, digestibility, energy and protein (HCC, 2008).

The protein provided with the help of forage is associated to the content of the nitrogen (Rinehart, 2008) and increased atmospheric carbon dioxide concentration have been shown to be linked with the decreased shoot nitrogen (Soussana et al., 1996; Ehleringer et al., 2002; IGER, 2003), along with the condensed vitamins (Soussana et al., 1996) and elevated soluble carbohydrate (Ehleringer et al., 2002; IGER, 2003).

The age of the plant also impacts the nutritional content of the forage. Talking about the older plants, older plants have less protein, more fiber, and lower digestibility (National Research Council, 1985; Rinehart, 2008).

In this way, increment in the rate of growth is linked with elevated atmospheric carbon dioxide concentration, which might permit the younger plants to be consumed, and in addition to that, it might help in order to elevate the protein as well as the soluble carbohydrates content for livestock (Rinehart, 2008).

Never the less, the exact effect of the sudden changes in the climate on grassland nutrient content remains unclear, there are several numbers of indications that decreases the protein in grazing might influence negatively on weight gain in cattle (Owensby et al 1996 in (Topp and Doyle, 1996b; Ehleringer et al., 2002; Craine et al., 2009). According to several numbers of published studies about the impact of climate change on the nutrition of grazing sheep appear lacking to date.

As stated by (Campbell et al., 2000), in the coming interval of time, the quality of the forage could decline in systems where the feed conversion productivity is limited by protein and can elevate or remain same in the systems where it is limited with the help of availability of the carbohydrate.

3.6. SHIFTING PLANT SPECIES COMPOSITION IN RESPONSE TO CLIMATE CHANGE STABILIZES GRASSLAND PRIMARY PRODUCTION

The structure ,as well as the functions of the alpine grassland ecosystems, which is consisting of their broad soil carbon stocks, are largely shaped with the help of the temperature. In particular, the Tibetan Plateau has experienced considerable warming over the time interval of around five decades, and this warming trend is anticipated to increase in the coming interval of time.

These kinds of sudden changes in climate will likely alter the species of the composition of the plant and net primary production (NPP). According to the integrated observation of more than three decades and monitoring with a manipulative experiment of the temperature and the precipitation, in order to explore the impacts of the alterations in the climate on the plant community structure as well as functions of the ecosystem.

First of all, long-term climate warming from the year of 1983 to the year of 2014, which took place without systematic alterations in the precipitation, led to higher grass abundance and lower sedge abundance but did bit impact aboveground net primary production.

Second is an experimental warming experiment that is conducted over the time period of four years, had no impact on any aspect of the net primary production. On the other hand, drought manipulation (which is reduced precipitation by almost half, that is fifty percent), shifted net primary production allocation belowground without impacting the total net primary production.

Figure 3.6. Shifting plant species composition in response to climate change stabilizes grassland primary production.

Source: Image by Flickr

Third, both drought treatments as well as experimental warming, aided by a meta-analysis at nine sites all across the plateau, increment in the grass abundance at the expense of the biomass of the sedges and forbs. This alteration in the functional group composition led to a deeper root system, which may have enabled the plant communities in order to accomplish more water. In this way, stabilize ecosystem primary production even with an altering climate.

Overall, according to the studies that demonstrate, changes in the species of the composition of the plant in response to the sudden changes in climate may have stabilized primary production in this high-elevation ecosystem, but it also produces a shift from aboveground to the belowground productivity.

Sudden changes in the climate will affect the structure and the function of the terrestrial ecosystem, more specifically in cold regions. The amount of the net primary production (NPP) in an ecosystem impacts the carbon feedback between the terrestrial biosphere and the atmosphere.

More and more experimental studies have explored how plant community composition and the primary productivity will respond to sudden changes in the climate. Never the less, several numbers of studies have done so. With respect to the long-term observations, in order to discover how alterations in the community composition that are triggered with the help of sudden changes in the climate can impact the functions of the ecosystem.

Alpine ecosystems are well thought out from the most sensitive ecosystems, with respect to the climate alterations. Near about sixty percent of the Tibetan Plateau is Alpine. Over the time period of around five decades, the annual temperature on the Tibetan Plateau has elevated very drastically with a rate of two times as compared to the global average.

In addition to this, data shows that, from the year 2000, the rate of warming can be increased even more as compared to the time span of the last four decades. Alpine grasslands are very vulnerable to warming, also they are diverse, and harbor large stocks of the carbon in the soil that, if they released, they can take into the direction towards positive feedback for climate warming.

The ubiquitous alpine grasslands on the Tibetan Plateau are very significant with respect to the livelihood of the native nomads and the potentially vulnerable to the sudden changes in the climate. Several numbers of approaches have been taken, in order to cope up with the effects of climate change on this high altitude, which is alpine ecosystems.

Both process-based ecosystem models and the long-term time series datasets of the Normalized Difference Vegetation Index (NDVI) recommends that increment in the vegetations over the time period of past several numbers of decades, never the less, with the pronounced geographical heterogeneity.

Even though, these kinds of approaches, while informative at large spatial scales, do not offer detailed fine-scale information or insights into the potential mechanisms. Long-term observations at the local plot scale have shown that net primary production in the alpine grasslands appeared to be very insensitive to the modifications in either precipitation or in temperature of the air from the year 1980 to the year 2000.

An observational methodology like the one that is taken in the study, cannot explicitly address the underlying mechanism that explains why the net primary production was unresponsive with respect to the sudden changes in the climate change. In the end, a number of manipulatives explaining experiments on the Tibetan Plateau, with the help of either open-top chambers or infrared heaters, have reported inconsistent outcomes, some showing the increment in the net primary production and the grass biomass. On the other hand, others are finding the decrement in the net primary production.In addition to this, never the less, experiments can be very useful in order to explain the mechanisms and providing the boundaries of the possible change. They cannot hold the legacy by the background warming and significantly, do not reflect reality.

Given these uncertainties, one vital way forward is to integrate the long-term observational studies with the experimental manipulations in order to examine the long-term patterns as well as the detailed mechanisms of the community and the ecosystem responses with respect to the climatic alterations.In the present interval of time, according to several numbers of studies, integration of around 32 years of observational monitoring of the alpine grassland ecosystems on the Tibetan Plateau with a manipulative experiment in order to discover the impacts of the climate change in the context of productivity of the grassland productivity and community composition.Particularly speaking, researchers have evaluated that, whether it is net primary production and the plant community composition were sensitive to the sudden changes in the climate in this high-elevation, cold systems, and whether the changes in the plant community composition amplified or stabilized changes in the productivity. In the end, it is also explored that the possible underlying mechanisms of the impacts of global warming on primary productivity.

3.7. CLIMATE CHANGE EFFECTS ON FORESTS AND GRASSLANDS

The average annual temperature in several regions of the United States has elevated over the time period of the last ten decades. Most of this warming has occurred from the time of 1970s. As a matter of fact, each and every decade from the time period of the 1980s, has set a new record for the warmest decade on the record.

The average temperature in the United States has been increased to more than two degrees Fahrenheit in the time period of the last five decades. Increment in the temperature is having extensive impacts on the forests as well as the grasslands. These impacts are consisting of precipitation patterns, insects, streamflow, snowpack, and invasive plant species, and influencing drought, wildfire, and heatwaves.

There are just some of the examples of the observed impacts of the sudden changes in the climate all across the United States, and most of the impacts are expected to lasts. And also, these impacts will be going to be elevated through the course of the century.

Emission of the greenhouse has been increasing more and more, and also, greenhouse gases are impacting the present conditions of the climate. The choices of humans make about land use, energy, and the development will impact the future amount of greenhouse gas emissions.

Figure 3.7. Climate change effects on forests and grasslands.

Source: Image by Geograph

In order to anticipate the future impacts of the sudden changes in the climate, several kinds of instances with respect to the emission of the greenhouse gases in the future also known as "pathways", that are used as inputs to the models of the climate. Each and every climate model, as an outcome, helps in stimulating the climate system in a different manner.

The future scenarios of the emission of the greenhouse gases are integrated with several kinds of climate models result in a range of temperatures and the precipitation outcomes from a warmer to much hotter, as well as drier to wetter. Generally, instances of higher emission of greenhouse gases lead to the anticipation of more severe changes. On the other hand, cases with lower emissions or pathways lead to the projection of the less severe changes.

The range of the emission scenarios and the outputs from the climate models represents most of the uncertainty in the future climate projections. The climate is certainly altering and will endure to alter or shift. But because the researcher still has some unknowns about the future. Humans have to work with a plausible range of future climates that integrate the variety of the emission pathways as well as the model projections.

3.7.1. Benefits/Stresses

Under the right set of situations or settings, some of the ecosystems may benefit from the alterations in the climate. These benefits can only last for a few decades, or it could be possible that it will last for much longer. Never the less, it is more likely that the sudden changes in the climate will worsen the present stressors in an ecosystem or add new stressors.

Every so often, these kinds of benefits and the stressors that have been identified, are a part of personal values and beliefs about how ecosystems should look and function.

It is very important to think about where the advantages and the stresses can take place. In this way, humans will get prepare in order to take advantage of the sudden changes in the climate and overcome challenges from climate change stressors.

3.7.2. Effects

There are several numbers of ways in order to describe the impacts of climate change on the forests and grasslands. In this topic of the chapter, the prime impacts on forests and the grasslands are integrated into four extensive categories. These categories are wildfire, water, disturbance, and vegetations.

Each and every category may consist of both advantages as well as stresses to the ecosystem's health and function. The impacts may not take place in the isolation, and are, in fact, are very likely to interact with each other.

3.7.3. Water/Precipitation

Several numbers of various regions all across the United States have already experienced alterations and modifications from the historical precipitations: some of the regions are receiving more precipitation and more heavy precipitation, and others are becoming drier. Precipitation in several numbers of places has become more variable, every so often confusing the clear trends. Alterations in the precipitation patterns and the amounts impact soil moisture, risk of wildfire, and level of groundwater.

Projections of the future precipitation show high variability, and every so often, several kinds of emissions scenarios depict opposite patterns in the rainfall through the course of the same season. Regions that are projected in order to experience more annual precipitation in the future may not necessarily benefit, because more of that precipitation is anticipates to fall through the course of intense storms.

This is already taking place in several numbers of regions, as an outcome in more frequent flooding and erosion. Some forests are experiencing longer dry periods between the heavy precipitation events. These dry periods can result in water stress in forests, and the heavy rains may not help in decreasing the water stress because more precipitation flows directly into the streams, in spite of filtering into the soil.

3.7.4. Water/Snow Cover

In the time period of last several numbers of years, snow cover across much of the United States has reduced in depth, extends over a smaller area, and melts sooner in the spring. Alterations in the snow cover and snowmelt have extensive impacts. Less snowpack integrated with the earlier melting that offers less insulation for plants and soil, exposing them to the frosts as well as freezing temperatures.

Early snowmelts also shift the timing of the runoff into the streams, with the large amount of flows taking place earlier, which is followed by diminished flows late in the growing season.

Low flows in the summer and fall can take into the direction that can lead to the water stress for the plants, that can take into the direction towards the elevated risk of insect infestations and the wildfire, and impact the soil processes. Lower flows late in summer also result in lower water volume and higher temperatures in streams, which may impact the population of fish in a negative manner.

3.7.5. Water/Flow

Altogether, the alterations in the precipitation patterns and snow cover impact the amount and the timing of the runoff as well as streamflow. Particularly speaking, rain-on-snow events, when rain falls on snow, are projected to become more frequent and intense as the temperature rises.

Warmer temperatures alterations through the course of winter precipitation from snow to rain, and the warmer rain filters into the snowpack, which produces melting, increment in the runoff and the danger of flooding as well as conditions of erosion.

The decrement in the summer seasonal flows and increment in the flash floods and short periods of high flows from more intense storms are anticipated impacts of the sudden changes in the climate through the course of the century.

3.7.6. Water/Sea-level Rise

Oceans absorb more than ninety percent of the extra heat in the atmosphere from the emission that is produced by humans. As the water heats up, it expands, contributing to the increment in sea level. Global sea levels have been elevated to near about 8 inches from the time period of the 1880s, and from the year 1992, the sea level all across the globe has been increased two times as fast as the long-term trend.

Figure 3.8. Increment in water / sea level rise.

Source: Image by Pixabay

By the end of the century, sea level all across the globe is anticipated to increase by one to four feet. This increment, along with the projections for more coastal storms, will elevate the danger of erosion, flooding events, storm surge, and impacting the infrastructure as well as the coastal ecosystem.

3.7.7. Vegetation/Growing Season

The frost-free season every so often is used to define the length of the growing season, has lengthened around the country since the time period of the 1980s, with the largest increase observed in the west. The first freeze is happening later in the fall, and the first thaw is happening earlier in the spring. The increment in the length of the frost-free and growing seasons is predicted to last in this century.In the earlier time period, snowmelt and springtime have led to earlier emergence of leaves, or leaf out, and earlier flowering in several numbers of plants. Longer warm periods mean that the plants have a longer time to prosper, which can elevate the productivity of the forest and grasslands if the plants have a sufficient amount of water as well as nutrients. This could be a very beneficial impact on the sudden changes in the climate.Never the less, there may also be some negative aspects to a larger growing season.

3.8. CONCLUSION

Climate change has adversely impacted the entire world. Along with human beings, flora and fauna have been heavily affected due to severe climate changes that are taking place in the entire world. This chapter discusses the impact of climate change on the grasslands. The grasslands hold a very important place in the entire ecosystem.

The grassland all over the world has experienced drastic changes due to severe climate change. Mainly due to the highly increased rate of CO_2, the grasslands have been highly impacted. But the impact of climate change on the grasslands is not as highlighted as forests. This is so because the impact of climate change is more clearly evident in the forests as compared to the grasslands. This is not because of the forest cover or the open fields in the grasslands.

The main reason why the forests more evidently display the impact change of climate is because of the storage of Carbon and its compounds in the leaves. On the other hand, the grasslands do not have high forests cover. Most of the Carbon that is stored in the grasslands are in the soil. The grasslands become capable of holding high content of Carbon is because of the long-grown roots of the grass and other plants inside the soil.

There are some grasslands that have been benefitted from severe climate change. Yet, the majority of the grasslands have been adversely impacted by severe changes in the climate. Thus, it is a very important focus on the vegetation and water flow to preserve and maintain the grasslands.

REFERENCES

1. Climate change connection (2015). Grasslands impacts | Climate Change Connection. [online] Climatechangeconnection.org. Available at: https://climatechangeconnection.org/impacts/ecosystems-impacts/grasslands-impacts/ [Accessed 19 Feb. 2020].
2. Conservation in a changing climate (2020). Climate Change Impacts to Grasslands. [online] Conservation in a changing climate. Available at: https://climatechange.lta.org/impacts-to-grasslands/ [Accessed 19 Feb. 2020].
3. Liu, H., Mi, Z., Lin, L., Wang, Y., Zhang, Z., Zhang, F., Wang, H., Liu, L., Zhu, B., Cao, G., Zhao, X., Sanders, N., Classen, A., Reich, P. and He, J. (2018). Shifting plant species composition in response to climate change stabilizes grassland primary production. Proceedings of the National Academy of Sciences, [online] 115(16), pp.4051-4056. Available at: https://www.pnas.org/content/pnas/115/16/4051.full.pdf [Accessed 19 Feb. 2020].
4. Parton, W., Ojima, D. and Schimel, D. (1994). Environmental change in grasslands: Assessment using models. Climatic Change, [online] 28(1-2), pp.111-141. Available at: https://link.springer.com/article/10.1007/BF01094103 [Accessed 19 Feb. 2020].
5. PARTON, W., SCURLOCK, J., OJIMA, D., SCHIMEL, D. and HALL, D. (1995). Impact of climate change on grassland production and soil carbon worldwide. Global Change Biology, [online] 1(1), pp.13-22. Available at: https://www.researchgate.net/publication/233714480_Impact_of_climate_change_on_grassland_production_and_soil_carbon_worldwide [Accessed 19 Feb. 2020].
6. Science net links (2020). Grasslands and Climate Change - Science NetLinks. [online] Sciencenetlinks.com. Available at: http://sciencenetlinks.com/lessons/grasslands-and-climate-change/ [Accessed 19 Feb. 2020].
7. SCRIPT Climate Change Effects on Forests and Grasslands: What You Need to Know. (n.d.). [ebook] Available at: https://www.fs.usda.gov/ccrc/sites/default/files/CC_Effects_Narration_Script1.pdf [Accessed 19 Feb. 2020].
8. van Oijen, M., Bellocchi, G. and Höglind, M. (2018). Effects of Climate Change on Grassland Biodiversity and Productivity: The Need for a Diversity of Models. Agronomy, [online] 8(2), p.14. Available

at: https://www.mdpi.com/2073-4395/8/2/14/pdf [Accessed 19 Feb. 2020].
9. Ze Han, Xiangzheng Deng, Xinliang Xu and Wei Song (2020). Grassland ecosystem responses to climate change and human activities within the Three-River Headwaters region of China. [online] Springer Nature. Available at: https://www.nature.com/articles/s41598-018-27150-5 [Accessed 19 Feb. 2020].
10. Hall & Scurlock 1991
11. Schimel et. Al. 1991
12. Long 1991
13. Owensby 1993
14. Long 1989
15. Eswaran 1993
16. Breymeyer & Melillo 1991
17. Parton 1993
18. Jenkinson 1991
19. Schlesinger 1990
20. Bailey 1989
21. Houghton 1990
22. Melillo 1982
23. Yonker 1988
24. IGER 2003
25. MAFF 2000
26. St. Claire 2009
27. Hopkins and Del Prado 2007
28. IGER 2004
29. Armstrong and Castle 1992
30. Topp and Doyle 1996
31. IGER 2003
32. Nowak 2004
33. ADAS 1997
34. Topp and Doyle 1996
35. Grime 2008

36. Soussana and Luscher 2007
37. Teyssonneyre 2002
38. Harmens 2004
39. Ehleringer 2002
40. HCC 2008
41. National Research Council 1985
42. Rinehart 2008
43. Topp and Doyle, 1996
44. Ehleringer 2002
45. Craine 2009
46. Campbell 2000

4

Grassland Management Techniques/ Methods

CONTENTS

The chapter of grassland management techniques and methods, explains the basic significance of the approaches and methodologies that are being used in grassland management. This chapter also includes the several kinds of grazing management techniques such as management intensive grazing fundamentals, meeting the nutritional needs of livestock, optimization of the forage base, selection of forage species, understanding grazing animals and their management, species response to grazing management, animal grazing efficacy, stocking rate and carrying capacity, and number of paddocks.

This chapter also addresses the several kinds of principles of grazing or grassland management, such as the timing of grazing, frequency of grazing, and intensity of grazing. This chapter also explains the basic concept of the herbivory, how plants respond to the grazing, what is increaser, decreaser, and invader. This chapter includes the impacts of grazing or defoliation such as intensity and the frequency of grazing, season of grazing, competition from other plants, and when is the process of grazing is good for plants? This chapter also provides a highlight of the production and conservation of grassland management.

4.1. INTRODUCTION

The branch of the agriculture, or plant husbandry are involved in the production of the hay, raw materials, and green fodder for the grass meal and other fodders on natural and sown hayfields and pastures. The science of the theoretical ideologies of grassland management.

Grassland management is worried about the production of the fodders both on meadows and on the other types of forage lands (steppes, bogs, and deserts). It plays an important role in the development of a stable fodder supply for public livestock production, with each and every passing year, hayfields and pastures account for a third of all fodder that is grown in the USSR.

The prime goal of the grassland management is to enhance the natural as well as hayfields and pastures, in order to create highly productive planted lands, and in order to make sure that the most effective application of all these lands. From the point of view of the organization and economic, grassland management is closely associated with the field cropping and animal husbandry.

The history of grassland management can be divided into four periods. Through the course of the first period, the original use of the forage lands, that is cattle grazed freely all across the year. (in winter, they found fodder under the snow.)

The animals were driven from one seasonal pasture to another. Through the course of the second period (11th and 12th centuries), grasslands were utilized as the hayfields and pastures. The cattle were grazed, and hay was laid in both as the main fodder for winter, when the cattle were stabled, and as a reserve for the period when the animals were maintained in the distant pastures.

The starting of the third period in Russian more or less overlapped with the emancipation of the serfs and the development of industrial capitalism, when a large amount of agricultural produce became very important to the cities. Experiment stations, bases and fields and the demonstration plots were planned in order to create methodologies of improving the natural forage lands. Never the less, these efforts were of practical value only for large estates.

The fourth (that is modern) period started after the Great October Socialist Revolution. In the year of 1920, land reclamation sub-divisions were planned in the provincial and district land agencies and were operated with the help of hydraulic engineers, grassland specialists, and land reclamation engineers.

These kinds of specialists have created land improvement systems on the natural forage lands, enhanced them with high pace, created drained lands, and sowed hayfields and the pastures. A great deal of work was done on the floodplains of the Iakhroma, Dubna, Viatka, and Moscow rivers.

With the beginning of the collectivization, stations of grassland reclamation and teams from machine and tractor stations (and, later, specialized mechanized reclamation and the grassland-reclamation stations and construction and assembly administrations) enhanced natural forage lands with the help of drainage and with the help of clearing the overgrown and weedy plots and reclaimed previously unusable land for the hayfields and pastures.

Some of the organizations that deal with the research and stuff, play a vital role in the grassland management with the help of finding out the efficient methodologies of improving and application of the natural grasslands. All of the basic processes of grassland management have been mechanized. Complicated cultivators, rotary tillers, sprinklers, power shovels, and aerators are now in use.

in the USSR, the natural grasslands cover the large areas, in the year of 1972, near about 374.6 million ha, out of which around 45.3 million were hayfields and around 329.3 million were pastures (that is excluding of reindeer pastures). Most of these lands are in the desert and semidesert zones, and several numbers of them produce low yields and are in need of developed measures. Enhanced and cultivated (sown) hayfields and the pastures inhabit near about ten million ha and are very productive (near about six thousand fodder units per ha of cultivated pasture).

Out of the country, grassland management is very necessary for the development of livestock production in Australia, the USA, Great Britain, New Zealand, the German Democratic Republic, the Netherlands, and the Federal Republic of Germany. Cultivated pastures are more particularly important in the folder production in Great Britain and the Netherlands, where they yield ten thousand fodder units or more per ha.

According to science, the management grassland creates the scientific bases and technology in order to enhance the natural hayfield and pastures and developed cultivated ones and develops the methods of applying them effectively. The natural-scientific foundation of grassland management is grassland science.

In several regions of Russia, the science of grassland management took shape at the turn of the 20th century. Experimental work was firstly done with the help of private initiative and later by the Department of Agriculture and with the help of zemstvos (which is a district and provincial assemblies).

The first experiment stations, fields, and plots for the research on the grasslands were founded early in the time period of the 20th century. In the year of 1917, well -known researcher named as V. R. Vil'iams and A. M. Dmitriev established a station near Moscow in order to study the forage areas and forage plants. The State Grasslands Institute (now the V. R. Vil'iams All-Union Fodder Research Institute) was founded in the year 1922.

From the establishment of the Soviet power, sectorial, zonal, and specialized research institutes and experiment stations have been organized in all of the natural zones of the country in order to perform the research on the fodder production and the grassland management. The All-Union Fodder Research Institue offers methodological guidance and coordinates the research.

According to the professor named as I. A. Stebut, who gave the first lectures in Russia on grassland management at the St. Petersburg Academy of Agriculture and Forestry (at present, known as K. A. Timiriazev Moscow

Agricultural Academy) from the time period of 1866 to the year of 1895. Vil'iams and Dmitriev succeeded him as lecturers.

Annual advance courses were developed with the help of the Riga Polytechnic Institute in the year of 1912 and the Moscow Agricultural Institute (at present known as the K. A. Timiriazev Moscow Agricultural Academy) in the year of 1913 in order to provide training to the professionals in grassland management. Persons with higher education and, more likely, with 2 or 3 years of practical experience were admitted to such types of courses.

Those who completed the courses were given the chance to travel to the Netherlands, Denmark, Sweden, Germany, and America and familiarize themselves with the foreign experience in the grassland management.

Professionals with the intermediate qualifications and masters in the grassland management and swamp cultivation were trained in the early 20th century in Chernigov, Viatka; Kiev, and Tver provinces. After the Great October Socialist Revolution, professionals were also trained in courses that are sponsored with the help of the State Grasslands Institute.

Figure 4.1. The basic significance of grassland management.

Source: Image by pexels

In the year of 1923, the K. A. Timiriazev Moscow Agricultural Academy started providing the training to the professionals with the highest qualification in the grassland management. In the later interval of time, these groups offered their knowledge to a number of other institutes. The names of the specialty were changed to the fodder production in the year of 1973.

The natural grasslands of the USSR were inventoried with the L. G. Ramenskii and I. A. Tsatsenkin. Near about four thousand and five hundred forage plants were identified and explained in three volumes.Kormovye rasteniia senokosov i pastbishch SSSR [Forage Plants of Hayfields and Pastures of the USSR], edited by I. V. Larin. Moscow-Leningrad, 1950-56).

Methodologies were established for the quick and radical enhancement of the natural grasslands, the creation of the cultivated hayfields and pastures in the several parts of the nations, and the breeding of the new grass diversities for the various kinds of hayfields and the pastures. Grassland management professionals face the task of developing the cultivated pastures with the yields of eight thousand to ten thousand fodder units.

In this connection, research is underway on the application of the large quantities of inorganic fertilizers and their impacts with respect to the quality of fodder and livestock production (that is, on the soil-plant-livestock-man system). Other objectives are the development of special-purpose pastures for young animals and the development of the special kinds of food products from the livestock (for instance, for the making of cheese).

The USSR is active in the international congresses on grassland management. In the year of 1956, the Lenin All-Union Academic of Agricultural Sciences organized the section for Grassland Management, Fodder Crops, and Animal Grazing in order to coordinate the work of the scientists of the country on these matters.

The section joined the European Grassland Federation in the year 1966. A national group was employed in the year of 1986 in order to coordinate the work of the Ministry of Agriculture of the USSR with that of the federation.

According to the research done by the grassland management in the USSR, reported that, in the industrial-scientific journal of the Ministry of Agriculture of the USSR, Korma ((Fodder; since 1972) and in the transactions of the Lenin All-Union Academy of Agricultural Sciences and the All-Union Fodder Research Institute

4.2. GRAZING MANAGEMENT: TECHNIQUES

One of the most critical decisions that are facing livestock producers in Illinois is how to utilize the forage resources on their farm in the best way. In typical cow-calf operation, forage supply more than eighty percent of the annual feed requirement. More than ninety percent of the diet of breeding ewes are forage.

The manner in which this valuable resource is utilized every so often determines the profitability of a livestock operation. Inappropriately, on several numbers of farms pasture and hay lands have received the least amount of management attention. In the recent interval of years, producer awareness of pasture management has been increasing. Several numbers of factors have brought this about.

Most importantly, the condition of the economy has forth producers in order to look to at their bottom line and this reflected the low returns from mismanagement of forages and pastures. And also, all across the globe, decrement in the feed grain prices and a new emphasis on the soil erosion and water has led several numbers of farmers with respect to a renewed interest in forage and grassland agriculture.

Figure 4.2. Grazing or grassland management technique: An illustration of practicing fire-fighting techniques.

Source: Image by Wikimedia Commons

In the end, the advancement in the technology of the pasture management and the improvement have made more intensive utilization of the forage resources more feasible with respect to the economy. Improvements or developments in the equipment, more particularly, fencing equipment and water equipment have also provided assistance in order to bring about an elevated interest in better grassland management.

4.2.1. Management Intensive Grazing Fundamentals

A sound management intensive grazing (MIG) system is built around three key factors:

1. Fulfilling the requirements of the nutrient on whatever class of livestock is involved;
2 Optimization of forage yield, quality and persistence; and
3. Use of suitable technology in order to develop a management system that is viable practically, economically as well as environmentally.

All of these factors are closely linked and should be well thought out from a total systems approach. For instance, if a producer chooses to be a stocker feeder cattle backgrounder, he or she will want to use high- quality forages in order to maximize the individual calves gain. He or she decides that alfalfa is the pasture crop that is needed to fulfill the expectation of the performance.

4.2.2. Meeting the Nutritional Needs of Livestock

The first deliberation in the context of designing a forage system is that it must fulfill the requirements of nutritional values of the class or classes of the livestock that takes participation. There is no class of ruminant livestock that cannot be raised or maintained on a hundred percent forage diet, having the expectation of "corn-fed-beef".

Some individuals every so often uses the grain supplementation as suitable for good forage management or even to remedy poor forage management. There are very real variations in the nutritional requirements of the several lasses of the livestock and there must be taken into consideration when setting up a grassland management program.

Most of the producers realize that the fluctuations in the nutritional demand of the stockers versus the dry cows or first -calf heifers with the suckling calf versus mature cows with calves but, people every so often oversee the more delicate variations of high milk- producing cows versus an average cow or straight bred calves versus crossbred calves.

4.2.3. Optimization of The Forage Base

At the time of choosing the species of plant, in order to use in a forage system, people try to decide on an optimal compromise of the factors of quality, yield, and persistence. Generally, people think of several numbers of various forages as having specific types of attributes.

People can easily visualize the endophyte- infected tall fescue as being very effective as well as persistent but of limited quality of bluegrass as being persistent and of high quality but a low yielding grass. Orchardgrass-alfalfa mix is very effective and high quality but may not be persistent under lax management or harsh environmental situations.

In this way, people must recognize the strengths as well as weaknesses of the players in order to integrate them into a balanced team. A good livestock producer should have as sound an understanding of the production characteristics of the forage that is available just as he understands the significance as well as the usefulness of the several numbers of the breed and the bloodlines of the cattle or sheep.

4.2.4. Selection of Forage Species

In several regions of Illinois, people are very fortunate that they have a great variety in adapted forage crops. People can extensively break these out into three classes that are based on plant biology:

1. cool -season grasses,
2. cool -season legumes, and
3. warm- season grasses.

Most of the sown pastures in Illinois are dominated through cool -season grasses, which is consisting of smooth bromegrass orchardgrass, Kentucky bluegrass, tall fescue, timothy, Reed canary grass, and perennial ryegrass, among others.

Figure 4.3. An illustration of Forage in the grazed field.

Source: Image by pixabay

The benefits of the inter seeding legumes in a cool -season grass pasture are well known and have been practiced with the help of several numbers of stockmen for number of years. Generally, users of legumes are red clover, birdsfoot trefoil, white clover, and lespedeza.

The third category is consisting of the most commonly used annual summer grasses ,for example, Sudan grass, sorghum and Sudan hybrids and pearl millet. In addition to this, the perennial species that were native to the American prairies have seen some renewed interest as forage crops.

These would emphasize switchgrass, Eastern Gama grass and Indian grass, and big bluestem. Actually, the individual species within each and every class may have quite different production trends that rely on the relative maturity, particular site adaptation, and production grazing management fertility can also change the curves.

4.2.5. Understanding Grazing Animals and Their Management

Animal care ,as well as their behavior, impacts the decisions that are taken on a daily basis about the movement and the animal regulation activities in the grazing systems. Animals behave first as individuals in their particular grazing selectivity, health, and reproductive cycle. On the other hand, practically it requires that people manage the animals as a herd or flock, which commonly develops and enhances the effectiveness of animal regulation.

There are inherent group behaviors that should be well thought out in the design and management of the grazing system. Cattle graze from near about eight hours to twelve hours per day and sheep from near about six hours to eight hours. Simultaneously, horses will graze up to fourteen hours to sixteen hours per day.

Cattle and sheep break this active grazing time into near about five to six separate intervals of grazing, with the time required for ruminating and resting between the intervals of grazing. The normal pattern of a horse is to graze non-stop for several numbers of hours, some rest, and then again grazing.

4.2.6. Species Response to Grazing Management

There are several numbers of desirable forage species that will persist under regular or nonstop, and uncontrolled grazing. Several examples are Kentucky bluegrass, tall fescue, and common white clover. These are the kinds of plants which would not be likely to show good feedback with respect to the execution of a controlled grazing program, never the less, some development or enhancement in the rate of use and animal output or acre could be predicted.

Upright growing legumes, for example, alfalfa and red clover, and grasses which increase the growing points early in the development, for example, bromegrass and Indiangrass, are more likely to respond very dramatically in order to regulate the grazing.

This difference is because of the two factors, that are mentioned below:

1. carbohydrate (CHO) balance in the plant, and
2. location of the developing seed head (meristem) relative to grazing height.

Low growing or prostate species maintain an appropriate substantial leaf area below the grazing height and then maintain a positive balance of carbohydrate (CHO) even under heavy grazing. The upright growing species maintain a very little leaf area below the grazing height, unless it is lightly grazed, and must depend upon stored carbon hydrate (CHO) for regrowth following defoliation.

With respect to the legumes, most of the carbohydrate (CHO) storage is in the root. On the other hand, with respect to grasses, it is mostly in the lower part of the stems. With frequent defoliation with the help of the grazing and no interval of rest to replenish CHO reserves, the plants will not thrive.

4.2.7. Animal Grazing Efficacy

If forage is too tall or too short, animals will be unable to consume a sufficient amount of grass through the course of the time they will graze each day in order to fulfill their requirements of nutrition. Grazing animals cannot greatly recompense for ineffective bite size with the help of grazing more hours through the course of a day.

Cattle graze with the help of bringing in forage with their tongue and tearing or shearing it off with the teeth on their lower jaw. The most effective forage height for a cow in order to graze ranges from 4 to 10 inches. It is a hard task for them in order to get appropriate bites if forage is shorter, and it necessitates much more time for them in order to get longer forage into a form, in such a way that they can swallow.

Horses have both upper as well as lower teeth and graze with the help of squeezing the forage. Sheep use upper lips along with lower teeth in order to graze nearly as close to the ground, just like horses do. Horses and sheep can both eat shorter forage more effectively as compared to cattle and their most effective forage height for grazing are ranges from near about 2 to 6 inches.

It is very hard for them to get adequate bites if the forage is very short. Sheep ,as well as horses, also have more struggle in order to graze the forage that is much taller as compared to their optimum.

4.2.8. Stocking Rate and Carrying Capacity

Potential stocking rate or carrying capacity is closely correlated with the total yield potential of a species but will fluctuate with the site adaptation, grazing system, fertility, and resistance with respect to the grazing damage. At the time of constructing a forage program, remember that it is only as good as its weakest element.

The species which provide high carrying capacity in the spring and very little in the summer are generally used in Illinois and must be worked around. Legumes or perennial and/or annual warm- season grasses are very important for a well-balanced forage system in order to meet the shortage of forage through the course of summers.

A good regulated grazing system will greatly extend the summer and fall productivity of cool -season grasses when compared to the regular grazed systems, more particularly with the species with the lower grazing resistance (for example, bromegrass, Orchardgrass).

Persistence and productivity of legumes will be elevated with the help of regulated grazing, offering fertility requirements are met, and this will translate into elevated carrying capacity. Tall fescue has been reported as a species having very good carrying capacity.

One main reason for this apparent impact is the low animal intake on the endophyte -infected fescue. It may be possible in order to maintain a high number of animals per acre on the infected fescue but the performance may be nil. When discussing the carrying capacity it should be linked to some kind of measure of the performance.

4.2.9. Number of Paddocks

There is no magic number of paddocks that one must -have for a successful regulated grazing system. The optimum number of paddocks will fluctuate with respect to the species because of the resistance to the grazing, economic potential, and regrowth habit.

The ideal system would have grazing animals move frequently to the fresh paddock. Never the less, this ideal is every so often is a very hard task to sell to several numbers of producers. The benefits of these kinds of system are consisting of minimal feed wastage, very high-quality feed each and every day, decrement in the parasite infestation, uniform grazing at a quick rate, and many more.

For starting out, critical grazer should have a minimum of eight to twelve paddocks. Most of the producers quickly see the advantages of more paddock and move in that direction. The prime goal of the increment in the numbers of the paddock is basically to increase the density of the stock in order to produce uniform grazing.

Figure 4.4. An illustration of paddock: Grazing management techniques.

Source: Image by Needpix.com

4.3. PRINCIPLES OF GRAZING MANAGEMENT

Organizing or regulating the rangelands is both arts as well as science. Scientific knowledge is integrated with the practical local experience in order to seek out the solutions to the particular kinds of issues that are related to the management. This topic of the chapter, will focus on how science can be utilized in order to regulate the grazing animals.

Grazing management involves how to manage or organize the grazing animals, forage plants, and soils interact in order to fulfill the requirement of the specific ecological as well as economic aims. An effective or efficient grazing manager needs to know that how plants grow and reproduce if he or she wants to understand how grazing animals impact the plants with the help of amount of the plant material removed, hoof action, as well as the other factors of grazing.

In addition to this, managers or executives must understand the grazing animal, more particularly, its nutritional requirements at crucial points of the production cycle as well as its behavior which is consisting of diet selection.

Managers are required to well thought out the factors that are mentioned below:

1. When should grazing occur? (timing)
2. How often should grazing occur? (frequency)
3. How much forage should be removed, or how much residual plant material should remain after grazing? (intensity)

How plants respond to grazing is found out with the help of timing, intensity, along with the frequency of grazing as well as the physiological and morphological aspects of plants and how these aspects will impact the growth of the plant before grazing and after grazing.

4.3.1. Timing of Grazing

Grazing managers should avoid the grazing a region at the same stage of the growth of the plant year after year. Alteration in the timing of grazing avoids repeated defoliation through the course of the crucial interval of time of the plant growth (that is associated with the Range Ecology), which benefits the plant. The most crucial growth stages are when the plants are started to grow in the spring or fall and more especially when the plants are started regrowth after the grazing.

The growth of the additional leave or regrowth of the leaves necessitates the energy from the plants. The plant also requires to store energy in order to grow in the future. The first two or three leaves that appear on a plant after a latent interval of time, utilize the energy that is stored from the previous growing season.

Plants require a sufficient amount of tissue of leaf in order to produce a sufficient amount of energy in order to meet both the growth as well as the storage needs.

If the grazing managers or grazing executives are not able to alter the season of grazing, in this way, plants have adequate leaf areas in order to fulfill their energy requirement for the growth of the plants and the storage. Then the executive or the managers are requiring to decrease the intensity or the frequency of the defoliation (grazing).

Grazing, through the course of the winter, when the plants are dormant, has few if any adverse physiological impacts on the plants unless grazing intensity and/or trampling are so simple they eliminate or harm the basal buds at the base of the plant that starts the growth of the following spring.

Moderate grazing through the course of the dormant season can help to decrease the build-up of dead plant material above the buds on the plant crowns. This can provide an advantage to the growth of the plant because more or higher quality sunlight reaches the buds and triggers them in order to grow.

In the end, the managers or the executives should avoid grazing when the soil moisture is very high for a long interval of time, and the reason behind this is that soil can be easily compacted with the help of trampling and become more susceptible to the erosion.

4.3.2. Frequency of Grazing

Grazing executive or the grazing managers should avoid the grazing plants too regularly through the course of a single growing season. If the plants are given a chance in order to regrow and refill their stored energy reserves after grazing, they can be grazed again or more through the course of a single growing season. Regrowth can be abundant under the following conditions:

1. when soil moisture is adequate for plant growth,
2. soil and air temperatures are optimum for plant growth,
3. soil fertility is good, and
4. grazed plants can produce new leaves after being grazed.

Never the less, in arid regions where the soil moisture for the plant growth is not sufficient for most of the year, grazing more than one time a year can be dangerous with respect to the plants.

If the grazing takes place very infrequently, the gathering of too much dead material will hinder the growth of the plants.

The growth potential of the plant is compromised due to the inadequate amount of sunlight reaches the buds of the plants in order to start the tiller growth. With respect to the several numbers of leaves, they do not receive the maximum amount of the sunlight.

Long intervals of no or limited grazing also can cause the nutritional quality of the plant to decline.

4.3.3. Intensity of Grazing

Grazing managers or grazing executives should avoid removing too much of the leaf area of the plant. Blades of the leaves are the prime area of the energy production for the plant. If the leaf area that remains after the grazing is very small or the growing points situated at the base of the leaf blade is removed, the plant may be not able to regrow and refill its energy reserves.

This is expected to take place if the moisture levels of the soil are low and not refilled in a short interval of time after grazing. Also, grazing managers or grazing executives should leave sufficient residual dry matter in order to avoid the erosion of soil and in order to safeguard the roots of the plants and stems, which bases from the extra amount of cold or heat.

Intense defoliation of the plants, more specifically if it takes place on a daily basis, it can decrease the leaf area of the plant for a long interval of time, and in such a way that the plant cannot store an adequate amount of energy in order to form the buds that are required for next year's growth and/ or sustain the buds across the long -dormant intervals.

4.4. HERBIVORY

The term Herbivory, or grazing, is a very strong ecological process. The process of herbivory can impact the amount and types of vegetation that is present on the landscape. Primarily, this takes place because herbivores graze selectively, choosing some plants or parts of plants over others, that helps in changing the competitive capability of the plants. Sustained heavy grazing can reduce palatable. On the other hand, preferred plants allowing the unpalatable, poisonous, as well as invasive plants in order to elevate in the community.

With the help of the regulation of the species of herbivore and the timing, intensity of grazing, and frequency, managers or executives can change a forb-dominated system to a grass-dominated system and vice versa.

"Targeted grazing" is the application of the grazing animals in order to achieve the particular kind of vegetation management goals with the help of strict control of the species of grazing animal, timing of grazing, and intensity as well as the frequency of grazing.

The absence of herbivory can also be a treasured tool with respect to the management of vegetation management. The absence of herbivory, whether, with the help of deferred grazing or complete rest periods, is designed in such a way that they can develop or enhance the forage stand.

Non-grazing intervals can be located to the particular pastures or worked into a planned rotation system. The advantage of planned, non-grazing relies on the time of year it is executed:

- Early spring — enhance leaf production by plants
- Spring — enhance plant re-growth when conditions are optimal
- Summer — allow for seed production
- Autumn — improve carbohydrate production and storage, particularly if summer dormant grasses had fall re-growth
- Yearlong — enable seedlings to establish, increase the vigor of preferred species, accumulate fine fuels for prescribed burning.

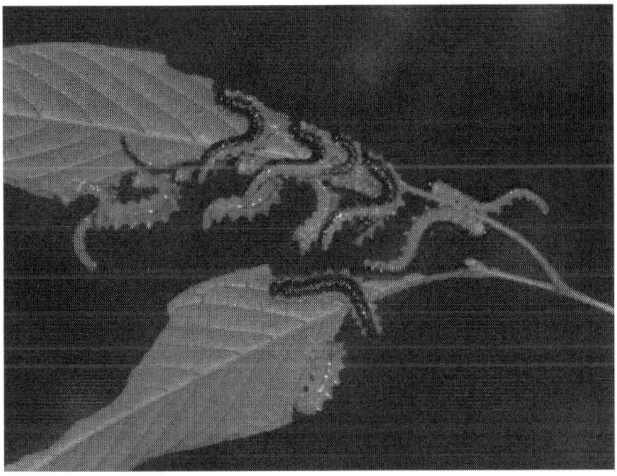

Figure 4.5. Herbivory: an illustration of herbivory caterpillar.

Source: Image by Wikimedia Commons

A regularly scheduled or occasional deferment can provide assistance to range types ,for example, mid-grass, semi-desert bunchgrass, sagebrush-grass, as well as mountain grasslands in order to elevate the forage plant reproduction, plant vigor, along with the general range condition.

The term "rest" is applied in grazing management in order to signify a full year (that is 12 months) of no grazing. Rest permits the plants to experience an entire growth cycle without being grazed. The advantage of rest is best realized in a special management situation, and these situations are mentioned below:

1. severe drought
2. following reseeding

3. providing fuel for prescribed burns
4. when critical site rehabilitation is required.

4.5. HOW PLANTS RESPOND TO GRAZING

On the other hand, the strategies in order to cope up with the grazing fluctuate greatly between the species of the plant. Generally speaking, plants are either avoid grazing or tolerate grazing. Plants avoid grazing with the help of diminishing their accessibility or the palatability to the herbivores.

Thorns, hairs, or the secondary compounds in some kinds of plants are examples of avoidance mechanisms. Plants that tolerate grazing have efficient mechanisms in order to ease the process of regrowth following defoliation. Usually, grasses are well thought out to have the highest grazing tolerance.

Generally, plants that withstand the grazing have one or more than one of the characteristics that are mentioned below:

1. Growing points are low, or elevation of growing points is delayed.
2. Plants have a high ratio of vegetative-to-reproductive shoots.
3. Apical meristems are activated, and new root growth is initiated following defoliation.

Range plants can be categorized with the help of how they respond to the grazing pressure. The amount of grazing pressure that a plant can withstand relies on how much it is preferred by the animals that depend on the grazing. The species of grazing animals can also impact how a plant responds to the grazing because several kinds of species have various dietary preferences as well as eating behaviors.

4.5.1. Decreaser

Decreaser plants are known as the first plants to die out under the condition of continued heavy grazing. These kinds of plants are declining due to they are either sought out with the help of grazing animals because of their high palatability.

In addition to this, they lack physiological characteristics that provide assistance in order to recover from the process of grazing. Plants that are highly preferred are the first to be grazed, and the animals can frequently graze these plants all across the growing season.

4.5.2. Increaser

Generally speaking, the increaser plants elevate their number as decreaser plants reducing. Several numbers of increaser plants can avoid damage to the grazing, and the reason behind this is, these plants grow close to the ground or are less palatable as compared to the decreasers.

Every so often, increasers also possess the physiological mechanisms that provide assistance in order to recover from the process of grazing. These plants should be supervised as they are a sign of high grazing pressure and can increment in the number as well as the abundance beyond what is necessary. If the overuse endures for a long interval of time, even increaser plants may decay in the community.

4.5.3. Invader

Invader plants are commonly known as the weedy plants. Invader plants become established because more and more needed plants have reduced because of the extra grazing process. A high proportion of invader plants in a community is generally a mark of overgrazing. Never the less, some of the noxious rangeland weeds, for example, leafy spurge along with the spotted knapweed, are having the potential of invading healthy rangeland plant communities.

4.6. EFFECTS OF GRAZING OR DEFOLIATION

The impacts of the grazing or any other kind of defoliation on a grass plant are directly connected to how the grasses grow. These kinds of impacts are relying on several numbers of aspects that are mentioned below:

4.6.1. Intensity and Frequency of Grazing

Grazing (such as trampling, defoliation, as well as other losses) intensity is the amount or percent of tissue of the plant that is removed with the help of grazing animals. The rate of recurrence denotes that how frequently a plant is grazed. Most perennial grass plants can tolerate either heavy grazing or frequent grazing but not both at the same interval of time.

When grazing removes, most of the leaf tissue, and when it takes place on a regular basis, the health of the plant is endangered and plants are more expected to suffer from conditions such as stress or even death. Everyday heavy grazing, as an outcome, in a plant with little leaf area for most of the growing season.

This has two opposing impacts, provided that it takes place on a daily across several numbers of years. First, when the area of the leaf is grazed 50 percent or more than fifty percent, on a daily basis, the root system will decline in size. Several roots mean the plant abstracts the less amount of water along with the nutrients from the soil and has not as much growth potential.

Figure 4.6. Effects of grazing or defoliation.

Source: Image by Pixabay

Second, the smaller leaf area is not that potential in order to produce adequate soluble carbohydrates (carbohydrates are the energy reserves) in order to develop buds, that are capable of enduring the winter dormant interval and develops the first one to three leaves the following spring. Without enough amount of stored energy in order to keep the basal buds thriving and yield the initial green leaves in spring, the plant will die.

4.6.2. Season of Grazing

At the time of year, when the animal grazes a plant impacts the ability of a plant in order to tolerate the defoliation. Grazing when plants are still in the vegetative stage, more specifically, early- to mid-growth, is comparatively less threatening, as the growing points are hardly ever detached and, normally, there is sufficient amount of soil moisture for the purpose of regrowth.

The plant can complete its growth cycle (until and unless it is grazed on a daily basis) and store an adequate amount of energy reserves in order to thrive through the course of the winter and produce the first few leaves the next spring.

Never the less, grazing when plants are in their early reproductive or "boot stage" of growth eradicates the apical ,as well as intercalary meristems (that are the growing points), are responsible for the growth of the plants.

When these meristems are eradicated, regrowth must start from the axillary buds at the base of the plant. This process is comparatively much slower, and also, it requires moisture as well as nutrients, generally at a time when the moisture of the soil is disappearing at a high pace. Growing conditions also impact the response to defoliation.

Plants withstand defoliation best when soil moisture and fertility are high. Drought or other environmental stress will reduce the amount of new leaf and tiller production following defoliation, which can hasten plant death.

4.6.3. Competition from Other Plants

Competition with the neighboring plants in the context of moisture, light, and nutrients can make stronger the impact of grazing. Plants can bear the process of grazing better when the neighboring plants are also defoliated. Never the less, Herbivores, graze selectively, every so often defoliating one plant and leaving other nearby plants ungrazed.

Frequent, selective grazing of a community's more palatable plants can, as an outcome, cause in their deterioration and rise of less palatable and/or less productive species. The replacement species are expected to be weeds, some of them are toxic with respect to the livestock.

4.6.4. When Is Grazing Good for Plants?

The process of grazing, when managed in a proper manner, it could be very beneficial to the plants and ecosystems in several ways. A lack of disturbance or defoliation, as an outcome, produces a build-up of dead plant material that "chokes" new growth of the plant. The result is overgrown decadent plants that yield less biomass (that is forage), have fewer seeds, and also, provide less nutritional value to the herbivores. Overgrown decadent plants are less resilient to disease, disturbance, and insects.

4.7. GRASSLAND MANAGEMENT: PRODUCTION AND CONSERVATION

With respect to India, the population of the livestock is near about 15 percent of the total livestock population all across the world, never the less, only 2 percent of the geographical area all across the globe. The forecast for the green and dry fodder prerequisite in India has been approximated at one thousand million tonnes and five hundred and ninety million tonnes by the year of 2010 A.D.

On the other hand, the present feed and fodder resources in the country can fulfill only around four percent of the prerequisite. The intensity of the grazing is very high, which is around twenty-six adult cattle unit / ha as against 0.8 ACU/ha in the emerging nations.

Figure 4.7. Grassland management: production and conservation.

Source: Image by Flickr

Near about 25 percent that is, 80 m ha of land of the country offers several levels of grazing to the animals. These regions are mostly natural grasslands which are consisting of barren and uncultivated land, grazing lands and culturable wastes, and permanent pastures.

The annual rainfall is inadequate in order to provide assistance to a flourishing growth of the trees. Never the less, grasslands are not constrained to the regions of low rainfall that is situated on hill slopes with patches of forests along with nala courses and depressions as in the shola region of South India.

The terai grassland of the foothills, grassland ecosystems form Himalayan pastures, semi-arid grasslands of Western and Central India, in the Shola forests of the Western Ghats, Nilgiri and Anaimalai ranges and in scrublands of the Deccan Plateau.

In this way, relying on the quantity of rain, there are short, medium and long grasses. Why manages grasslands?

Environmental Benefits

1. Prevents grassland from turning into brushland.
2. Maintains or develops the erosion of soil, quality of the water, quality of soil and carbon sequestration benefits of existing grassland.
3. Protects restored habitat for several numbers of plants as well as animals are important that is consisting of pheasant, song-birds, ducks, and endangered species. Grazing lands offer the forage that is the heart of the beef, goat, sheep, dairy and horse operations of the area.

Practical Benefits

1. Enhances the quality and yield of forage, biomass or native seed production.
2. Keeps unwanted species from getting established and becoming a nuisance.
3. Offer chances for birding, hunting as well as wildlife watching.
4. Keep grass-based conservation practices working properly.
5. May provide assistance to nearby crop health with the help of providing long-term habitat for animals that eat insect pests, along with the habitat for pollinator species that several numbers of crops depend upon, for example, bees.
6. Provides assistance in preventing the gully formation that necessitates the exclusive corrective measures.
7. Provides a low-cost alternative to growing crops on marginal land.

4.7.1. Areas of the World Covered By Grasslands

There are two common terms in grassland ecology. These terms are "Range lands" and "pastures".

The rangelands suggest to large, naturally vegetated, mostly unfenced grasslands of low rainfall areas that are grazed with the help of domesticated animals as well as wildlife.

Pastures are the enhanced grazing lands that receive abundant rainfall and are even irrigated, fenced, and commonly fertilized and seeded for forage plants.

Grasslands, more particularly, those with climatic climax vegetation, are a distinct ecosystem. The steppes, prairies, range pampas, most of the savannahs, etc., in several parts of the world, have climax vegetation.

4.7.2. Grasslands in India

The area under permanent pastures and grasslands – 13 m.ha Area degraded through special problems: Waterlogged – 11.60 m.ha Alkali soil - 04.50 m.ha Saline soil - 05.50 m.ha Acid soil - 25.00 m.ha Ravines & Gullies – 03.97 m.ha Shifting cultivation – 04.91 m.ha Riverine & torrents – 02.73 m.ha Flood prone area - 40.00 m.ha

The grassland survey carried out by the ICAR (Indian Grassland and Fodder Research Institute, Jhansi) Total land resource – 328.720 m ha. The available land resources for grazing Land resource Percent Forest cover 21.05 Permanent pasture 3.94 Miscellaneous tree crops and grooves 1.14 Cultivable wasteland 5.08 Fallow lands 7.52 Total 38.73

4.8. CONCLUSION

In the modern world, there has been a need for implementing modern grassland management methods and techniques. This is so because livestock dwells in grasslands. Due to changing climate, the grasslands have been heavily impacted. This makes it very hard for the livestock to compete for their needs and thus the sustainability of the livestock is in danger. Thus, it is very important to understand the need for managing the grasslands. One of the most efficient techniques for maintaining the grasslands is Grazing techniques. In the grazing technique, the main method is Management Intensive Grazing Farming. This method helps in efficiently managing the grassland resources. This method focuses on increasing the forage yield. This eventually helps in completing the nutritional requirements of the livestock.

There are various principles that have an impact on the process of Management Intensive Grazing Farming. It is very necessary to understand that Herbivory plays a very important role in grassland management. Herbivory is one of the strongest ecological processes. This is so because a large population of animals is herbivore and they depend on the grasslands for their survival. With the help of Grassland management techniques, it has been observed that various aspects have to control the grazing of animals in the grasslands. Grasslands can be effectively managed with the help of Targeted grazing. Targeted grazing helps in getting a particular species of plants grazed by the animals while conserving the other plant species. This proves to be very helpful as the plants that show rapid early growth are grazed by the animals. This prevents the animals from grazing plants with a slow growth ratc. Moreover, the animals receive proper nutrition from the plants that show rapid growth rates. So, there are a number of factors such as the timing of grazing, growth of the plants and the frequency of grazing that are very important for grassland management. If grazing is not effectively controlled and grasslands are not managed, then it would result in defoliation of the grasslands.

REFERENCES

1. da Silveira Pontes, L., Maire, V., Schellberg, J. and Louault, F. (2015). Grass strategies and grassland community responses to environmental drivers: a review. Agronomy for Sustainable Development, [online] 35(4), pp.1297-1318. Available at: https://link.springer.com/article/10.1007/s13593-015-0314-1 [Accessed 24 Feb. 2020].
2. Ed Ballard (2005). Grazing Management: Principles and Techniques - Pasture. [online] Livestocktrail.illinois.edu. Available at: http://livestocktrail.illinois.edu/pasturenet/paperDisplay.cfm?ContentID=8155 [Accessed 24 Feb. 2020].
3. Mr. Vikas kumar (2015). Grassland Management: production and conservation. [online] Slideshare.net. Available at: https://www.slideshare.net/vkskumar49/grassland-management-production-and-conservation [Accessed 24 Feb. 2020].
4. N. S. KONIUSHKOV (2010). The Great Soviet Encyclopedia: Grassland Management. [online] TheFreeDictionary.com. Available at: https://encyclopedia2.thefreedictionary.com/Grassland+Management [Accessed 24 Feb. 2020].
5. Ruyle, G. and Young, D. (2010). Introduction to Grazing Management | Global Rangelands. [online] Globalrangelands.org. Available at: https://globalrangelands.org/topics/uses-range-and-pasture-lands/introduction-grazing-management [Accessed 24 Feb. 2020].
6. I. V. Larin. Moscow-Leningrad, 1950-56
7. Fodder 1972

5

Grassland Management Laws and Policies

CONTENTS

This chapter named 'Grassland Management Laws and Policies' basically includes different laws and policies which are globally applicable for the grasslands and rangelands. In the first half of the chapter, it provides insights about the policy framework including its purpose, goals, objectives, key objectives and the guiding principles. In the second half, it mentions the policies which are planned and implemented by the European Union and FAO. This chapter contains two examples of countries that have planned specified policies and regulations and have implemented for the betterment of the greenness of their countries.

5.1. INTRODUCTION

Laws and policies are highly required in our society. These are important in terms that they aid to maintain the law and order in different societies. It also helps in shaping the social as well as political aspects of society.

However, laws and policies can be interrelated. Definitely, they are two different terms for different purposes. No country can go forward without having particular laws and policies.

A policy is something that highlights what a government is going to do and what it can attain for society as a whole. The term "policy" means what a government does not intend to do. In addition, it evolves the principles that are required for attaining specific goals. Policies are just the documents and not law, although these policies can lead to new laws.

Laws can be defined as set principles, standards and procedures that should be followed in society. Law is specifically made for the implementation of justice in society.

There are different kinds of laws that are framed in the past, such as civil laws, criminal laws, and international laws. Though, a law is framed for bringing justice to society. The policy is framed for attaining certain goals.

Laws are made for the people, while on the other hand, policies are made in the name of people. Policies can be called a set of rules that effectively guide any organization or any government. Laws are basically administered by the courts. Laws are enforceable at the places where the policies comply.

A law is formal as it is known as a system of rules and guidelines. In turn, these rules and guidelines are derived for the equity and welfare in society. On the other hand, a policy is informal as it is only a statement or a document of what exactly is projected to be done in the future.

Understanding the difference between laws and policies, and the intersections between the two can be helpful in understanding what type of career path you would like to take or any kind of additional education, for example, law school, a Master's in Public Administration, or both, are necessary.

Figure 5.1. Laws, Policies and Regulations is the participation of government in the management.

Source: Image by Pixabay

5.2. LAWS, REGULATIONS, AND POLICIES

5.2.1. Law

Let us take an example of the United States to know about laws related to grassland management. The authority in the US to protect, manage and administer the National Forest System, and several other land areas under Forest Service administration for range management purposes, is found in the acts given below:

- **Organic Administration Act of 1897:** This act authorizes the president of the country to change or modify any of the instruments to create a National Forest. It states that no National Forest can be established except for the improvement as well as protection of the forests within its boundaries. This is mainly for the purpose of securing pleasant conditions of water flows and also to provide a continuous supply of timber for the effective use of citizens of the US. It authorizes the secretary of agriculture to circulate rules and regulations to control the use as well as the occupancy of the National Forests.

- **The Bankhead-Jones Farm Tenant Act of 1937:** This act directs the secretary of agriculture to develop a program for the conservation of land and its effective utilization to correct instabilities in land use and hence, assist these things as the control of soil erosion, reforestation, conservation of natural resources and protection of wildlife.
- **Sustained Yield Forest Management Act of 1944:** It is basically an agreement between different secretaries of agriculture and the interior, or with some other federal agencies that have the authority over the forest land. Each of the secretaries is further approved in their discretion to enter into cooperative agreements with the other secretary. It can be with any federal agency having control over federally owned or administered forest land, or with any local or state agency having control over publicly owned or administered forest land.
- **Organic Act of 1944:** This act provided the establishment as well as protection of the water rights. These are authorized to be used for the expenditure by the forest service that may be necessary for the analysis as well as the establishment of the water rights, comprising the purchase of lands or interests in lands or rights-of-way for effective use and protection of water rights that are important and beneficial in relation to the public and administrative use of the national forests.
- **Wild Horse Protection Act of 1959:** This act prohibited the utilization of a motor vehicle to hunt, for killing or capturing any mare, wild horse, burro or colt running at large on the public lands. It also prohibited the pollution of watering holes majorly on the public lands for the drive of killing, wounding, trapping, or maiming any of these animals.
- **Granger-Thye Act of 19950:** The act authorized a wide range of improvements from appropriate funds and allowed the forest service to allow grazing advisory boards and also to issue grazing permits for periods not more than ten years.
- **The Multiple Use Sustained Yield Act of 1960:** This act establishes the policy as well as the purpose of the national forests to provide for multiple uses and persistent yield of the products and services.

- **Forest and Rangeland Renewable Resources Planning Act of 1974:** It authorizes long-range planning by the United States Forest Service to guarantee the future supply of forest resources along with maintaining a quality environment. It needs a renewable resource assessment and a forest service plan to be prepared every five and ten years to plan as well as prepare for the future of natural resources.
- **The Wilderness Act of 1964:** It established the National Wilderness Preservation System.
- **The National Historic Preservation Act of 1966:** It secures the protection of archaeological resources as well as sites on public and Indian lands.
- **The National Environmental Policy Act of January 1, 1970:** It directs all federal agencies to consider the key environmental impacts of the proposed federal actions and established the Council on Environmental Quality.
- **National Forest Management Act of October 22, 1976:** The National Forest Management Act modernized, expanded and amended the Forest and Rangeland Renewable Resources Planning Act of 1974, which was basically called for the effective management of renewable resources on National Forest lands. It is the main statute that governs the administration of National Forests.
- **Wild Horses and Burros Act of 1971** - This act helps in protecting wild free-roaming horses as well as burros from capture, harassment, branding or death. This act states that it is important to consider them in the area where an integral part of the natural system of the public lands is found presently.
- **The Clean Water Act of 1972** -This act named The Clean Water Act sets the basic structure in order to regulate the discharges of pollutants to waters of the countries like the United States.
- **The Endangered Species Act of 1973** -this acts helps in protecting the species of animal and the plant which are in danger of extinction (that is endangered) currently and the species which can be endangered in the upcoming future (which is basically referred to as threatened). It helps in conserving the ecosystems on which the threatened and endangered species depends such as fish, wildlife, and plants, both through Federal action as well as by encouraging the establishment of state programs.

- **Forest and Range Renewable Resources Planning Act of 1974** -This act named Forest and Range Renewable Resources Planning Act which was found in the year 1974, establishes public land policy as well as guidelines in order to manage, protect, develop and enhance the public lands.
- **Federal Land Policy and Management Act of 1976** -This act which was established in the year 1976, states that the public lands can be managed in such a manner that it will protect the quality of scientific, historical, scenic, ecological, air, environmental and atmospheric, water resource, and archeological values.

The management can be done wherever required, and eventually, this will preserve as well as protect some specific public lands in their natural condition. This ,in turn, will provide food as well as habitat for fish and wildlife and domestic animals.

Furthermore, this will provide for outdoor recreation and human occupancy as well as its use. This act also states that countries such as the United States shall receive fair market value for using the public lands and their resources until unless it is otherwise provided for by law.

- **Public Rangelands Improvement Act of 1978** -This act came into existence in the year 1978 and this establishes and reaffirms the national policy as well as the commitment to inventory and furthermore, helps in identifying the current public rangeland conditions and trends.

It helps in managing, maintaining and improving the condition of public rangelands so that, they become much productive and feasible for all rangeland values considering the management objectives and the planning process for land use.

This act charges a fee so as to use public grazing which is equitable; it also, continue the policy that is for the protection of wild free-roaming horses and burros from capture, branding, harassment, or death. At the same time, it facilitates the removal and disposal of excess wild free-roaming horses as well as burros which pose a threat to themselves and their habitat and also, to other rangeland values.

- **The Rescission Act of 1995 (Public Law 104-19)** - This act came into existence in the year 1995 and this required each National Forest to establish as well as adhere to a schedule for completing the NEPA analysis and decisions that are for all grazing allotments within 15 years.

5.2.2. Regulations

There are a number of examples for the policy which relates to the range resources as well as coordination of the range activities of the agencies such as USDA and other executive agencies, organization and individuals and some of the examples are mentioned below:

- Departmental Regulation, Number 9500-5 - dated December 15, 1983; Subject: Policy on Range.
- Executive Order 12548 - Grazing fees, 1986 - provide for the establishment of appropriate fees for the grazing of domestic livestock on public rangelands.
- Secretary's Administrative Order of August 1963, Administration of Lands Under Title III of the Bankhead-Jones Farm Tenant Act; Establishment of National Grasslands.

5.2.3. Policies

Some of the examples of the policies have been mentioned below which are followed by some of the countries such as US:

- National Grasslands, A Primer (1997) - This is a document that helps in identifying and interpreting the laws as well as the regulations which are applicable to the administration of the National Grasslands.
- Forest Service Rangeland Management Directives - These directives collectively make up the Forest Service's Rangeland Management Manuals and Handbooks.

5.3. SUSTAINABLE MANAGEMENT OF RANGELAND RESOURCES

Rangelands resources are said to be renewable. But there are a number of factors because of which severe degradation of the resources is being done and these factors include overexploitation, climatic changes, cultivation of rangelands, uncontrolled wildfires and poor management.

Such kind of situation leads to loss of palatable species and voluntary replacement that is by less palatable and invader species in order to fill the void, destruction of the microenvironment in which some other species may flourish, loss of vegetative cover which acts as the protective cover, decrease in biomass production and accelerated soil erosion.

Therefore, it is important to make sure that the management of rangelands should be proper and systematic for optimum plant and animal productivity by considering the mechanisms that allow for equitable access as well as the participation of a number of stakeholders.

It is important for the government to make sure that the sustainable management of range resources must be done in order to curb land as well as vegetation degradation and this can be done by the development and implementation of appropriate strategies for the rehabilitation of lost rangeland resources.

This can also be done by making sure equitable access as well as the participation of stakeholders in the sustainable development of rangeland resources.

5.3.1. Objectives

There are a number of objectives of sustainable development which must be done in order to manage the rangeland resources. Such objectives include:

- Promoting the integrated approach to planning as well as management of rangeland resources,
- Developing the strategies in order to properly manage the rangeland resources,
- Promoting the effective stakeholder's participation in the planning as well as the implementation of rangeland management programs, and
- Developing the appropriate policy and strategies for rehabilitation and possible restoration of lost rangeland resources.

5.3.2. Strategies

There can be many strategies which can be implemented while managing the rangeland resources sustainably and this can include:

- Foster integrated planning for sustainable management of range resources,
- Review and strengthen existing guidelines as well as regulations on grazing control management,
- Develop and facilitate the implementation of grazing management plans, which in turn can also promote the livestock grazing system that recognizes rest and rotation and variations of deferment of

grazing areas,

- Strengthen and resuscitate existing Range Management Associations and promote the establishment of a new structure where required,
- Restore degraded rangeland which can be done ecologically sound methods by stakeholders including communities and herders,
- Promote fodder production and storage for stall feeding programs,
- Explore options and benefits for the provision of environmental services on the rangelands,
- Work with the ministry which is responsible for environment, advocate for the resuscitation of The National Environment Council
- Establish as appropriate an environment coordinating body for coordination as well as co-operation among appropriate stakeholders which include local authorities, private sector and other organizations that are engaged in the programs related to the environmental protection, and
- Provide appropriate information and education and also, to integrate indigenous knowledge to the various levels of stakeholders on proper planning, management and implementation of rangeland resources management programs.

Example 1: Sweden

Sweden is a part of European Union (EU). Therefore, the Sweden forage and grassland policies are EU policies which have been adapted by locals by the Swedish Board of Agriculture (in the year 2015). There are many policies and subsidies which accompany them. These policies have recently been updated. Therefore, the positive or negative results of these policies are not apparent yet.

Generally, there are two types of grasslands in Sweden. One of them is short - term fields which are harvested for hay, silage and pasture which provide almost all the forage in the country. The other one is the semi - natural grassland which in earlier times have played the major role for forage as well as grazing. In today's time, they are mainly used for the purpose of extensive grazing. There is no true rangeland in Sweden. Possibly, there can be exception of land which is used for grazing reindeers in the North.

In the past, the subsidies given to farmers have mainly focused on keeping semi-natural grassland in use as much as possible because of the fact that they are the most biodiverse ecosystems in Sweden. Thesegrasslands cannot be fertilized and it is important that it must be grazed yearly for a farmer so as to receive USD 116 per ha.

There can be increase in the amount for more specific management, while mimicking those which were used in the earlier times. Currently, there are also subsidies for all kinds of cattle whose age is more than 1 year.
In today's time, there is a scheme which has the aim to increase the greenness of a farm. It seems to be complicated to calculate how much will be given to a particular farm, especially considering those which include clovers (Swedish Board Ag., 2015).

5.4. POLICY FRAMEWORK

5.4.1. Purpose

There are countries such as Lesotho, whose government identified the need to revise the current policy. This was done so as to develop a more comprehensive framework so that the poor state of the rangelands can be addressed. There is a fast deterioration of rangeland resources which is still continuing, like by the loss of valuable flora and fauna, drying of wetlands and other negative changes in the ecosystems. This fast deterioration has provided a compelling case of urgent as well as appropriate policy intervention.

Therefore, the policy intends to provide guidance so as to develop effective strategies in order to combat land and vegetation degradation and also, to motivate improved legislation and effective implementation.

5.4.2. Goal

The main aim or goal of the policy is to attain sustainable development and further, management of rangeland resources considering the enhanced biodiversity, optimum productivity and improved livelihoods of the present generation as well as the future ones.

5.4.3. Objectives

There are certain key objectives of the policy and these objectives include:

- Developing and implementing the strategies which are efficient and effective in order to avert land and vegetation degradation,
- Raising public awareness and furthermore, promoting the community and wider stakeholders' active participation in rangeland resources management,
- Improving as well as maintaining the productivity of rangeland

resources at an optimum level for the promotion of ecosystems balance,

- Conserving and increasing the availability of native plant species for economic, social and cultural utilization,
- Rehabilitating and improving the quality of rangeland for the enhancement of productivity of livestock and wildlife habitat,
- Enhancing the aesthetic beauty of the landscape to increase opportunities for sustainable recreation and ecotourism,
- Promotion of disaster, risk reduction, gender equity, as well as HIV and AIDS while considering it in range resources management,
- Protection of water resources as well as improvement in the water quality and yield, and
- Improvement in income opportunities and also, in the quality of life of the rural communities.

5.4.4. Key Policy

There are areas which are based on the policy issues, so, the main focus areas have been identified which include:

- Sustainable management of rangeland resources,
- Maintenance and protection of wetland areas,
- Conservation of biodiversity and maintenance of the ecosystem,
- Socio-economic dimensions, and
- Rangelands monitoring and research.

Figure 5.2. Representation of Policy.

Source: Image by The Blue Diamond Gallery

5.4.5. Guiding Principles

There are certain guiding principles in order to make sure that successful implementation of the policy directives have been done and these principles include:

- Harmonization of public sector policies and strategies,
- Effective and efficient governance system at the local level,
- Community participation and involvement,
- Sound technical support and advice,
- Gender equality and mainstreaming,
- Cooperation, collaboration and partnership among key stakeholders, and
- Effective policy and legislation enforcement.

Example 2: Japan

There are four main policies in the country Japan relating to the agriculture. These agricultural policies affect grassland agricultural production in Japan. Each of the policies have positive as well as negative implications for the farmers, the society and the environment as a whole.

Policy 1: *The cost of Forage Production is increased and the cost of Animal Management is also decreased.* In the policy, the first part focuses on enhancing forage production (both quantity as well as quality) which can be done by encouraging renovation or improvement of grasslands which can further be done by using techniques such as introduction of high - performance forage species and cultivars, and control of noxious weeds.

In the policy, the second part encourages increased stocking of dairy as well as beef cows on pastures versus indoor feeding. These policies prove to be beneficial to primary producers which is by increasing the costs of forage and animal production and decreasing animal management (MAFF, 2015).

The first part of the policy is mainly concerned with the renovation of existing grasslands and the reason is the high cost involved while developing new grasslands (such as conversion of forests into grasslands).

Therefore, the first part of the policy is required to have a minimal detrimental effect on the environment. However, the policy can have an adverse effect on the soil conservation (risk of erosion) and water quality (i.e. - manure and urine) which in turn depends on the stocking density. The reason is that the policy can encourage increased pasturing of livestock. Despite this, the policy is acceptable to society. This is because it helps in encouraging the production and utilization of domestic resources as well as improved maintenance of "green space" (MAFF, 2015).

Policy 2: *Measures for the Meso-mountainous Region and Abandoned Farmlands.* The mesomountainous region (which include intermediary area between plains and mountains) is responsible for around 73 percent of land area of Japan.

The issue that is most critical in this region is the abandonment of farmlands which is because of the aging of primary producers and labor shortages.
This policy helps in encouraging the use of land resources in the mesomountainous region for forage as well as livestock production which can be done by a number of different management techniques which further depends on the natural and social conditions of the land (such as conversion of abandoned farmlands to grazing land). It can be considered beneficial for primary producers by restoring abandoned farmland for the agricultural production.
The policy is also considered to be beneficial to the environment. The reason is that there will be a decrease in the area of abandoned farmlands which eventually reduce the spread of weeds, pests and wildlife. This damages the agricultural crops and the environment.
This policy said to be is beneficial to society. Because the expectation is that it can lead to the conservation of national land as well as the revitalization of the region (MAFF, 2015).
Policy 3: *Measures for Preventing Damage by Non-Native Species.* In Japan, there are many forage species which are used and most of them are non-native. These further introduced species. Therefore, the policy demands that these species is required to be maintained in a cultivated system as well as it is important to manage so that it cannot cause damage to the ecosystem. This policy in turn affects the primary producers in a negative manner and this happens because it increases the production cost and management requirements, but also, it provides a positive benefit to the environment and society (MOE, 2015).
Policy 4: *Intensification of the Prevention, Monitoring and Control of Animal Diseases.* This policy is considered to be definitely beneficial for the primary producers. The reason is that it decreases the risk of animal loss due to diseases. On the other hand, it can also increase the costs for producers by utilizing public pastures. Because urban people also use them for recreational purposes. The opportunity for the urban people to have contact with farm animals tend to decrease which in turn can be unfavorable to society (MAFF, 2015).

5.5. ECOLOGICAL SUSTAINABILITY AS THE FOUNDATION OF NATIONAL FOREST STEWARDSHIP

For the planning, the guiding star is sustainability. Similar to other overarching national objectives, sustainability is considered to be broadly aspirational and can be difficult in order to define in concrete terms

Especially considering the increased human pressures on the national forests and grasslands, it becomes even more important that planning and management will start with a central tenet of sustainability. It is important that our use today does not impair the functioning of ecological processes and the ability of these natural resources is considered to be important which in turn contributes in the future economically and socially.

There is a suite of laws that call for ecological sustainability, most of the time this is in terms of native - species diversity and ecological processes. There is an act that is named as Endangered Species Act which calls for federal agencies in order to undertake all possible means for conserving the native species and the ecosystems upon which they depend.

One more act named the National Forest Management Act which calls for the maintenance of the diversity of plant and animal communities so as to meet multiple - use objectives. While implementing the Act, these objectives in the existing regulations have been stated as providing habitat for the maintenance of the viability of vertebrate species and the protection of streams and watersheds.

An act named as the Clean Water Act calls for protecting the physical, chemical as well as biological integrity of the nation's waters. The act named as the Multiple - Use Sustained - Yield Act calls in order to make sure that multiple use and sustained yield do not impair the productivity of the land.

The act named the Clean Air Act calls for protection of the nation's air. Individually and collectively, the laws related to the environment express a profound commitment for protecting the native species and to our air, water and soil.

The thrust is clear, while considerable discretion is left in interpreting such laws. Considering the Northwest and throughout the country, the letter and spirit of these laws were ignored time after time, and this has led to lawsuits that have overturned long - term plans for the national forests and grasslands. This ,in turn, frustrates the people who have put a number of hours into them.

Along with this, the abrupt changes most of the time follow court decisions and these can be disconcerting to communities and economies. According to the committee, it would be better for the Forest Service to consider its responsibilities for ecological sustainability as a part of planning instead of waiting for lawsuits to force the issue. Scientific results and common-sense focuses on the necessity of protecting these systems so that they continue providing benefits to society. This was along with the suite of environmental laws calling for the protection of ecological systems.

There are points from the National Forest System which need to be considered. This suggests that the conservation of ecological systems cannot be ignored. Let's take an example. A number of concerns are there about the effect of declining water clarity on tourism in Lake Tahoe and these concerns have resulted in an intensive and expensive effort so as to reverse this trend.

The watersheds supply the municipal water for a number of communities and the cost of replacing the watersheds has caused an increase in the protection of these lands.

Once the ecological systems are pushed to the edge, the costs of recovery can be considered as astronomical and the ability to apply the adaptive management can considerably be compromised.

The Committee recommends that ecological sustainability provide a foundation upon which the management for national forests and grasslands can contribute to economic and social sustainability. This finding does not mean that the Forest Service is expected to maximize the protection of plant and animal species and environmental protection to the exclusion of other human values and uses.

Rather, it means that planning for the multiple use and sustained yield of the resources of national forests and grasslands should operate within a baseline level of ensuring the sustainability of ecological systems and native species. Such use would avoid impairing the functioning of ecological processes and the ability of these natural resources to contribute economically and socially in the future.

Setting ecological sustainability as a key goal acknowledges that ecological systems provide many outputs that humans require to sustain themselves as living, biological organisms. That is, human health and the integrity of ecological systems are inseparable objectives. Humans are "a part of" not "apart from" their environment.

Choices in management still exist, and the level of risk to take is a policy choice. Further, the human values, needs, uses, and ecological conditions of each locality will change with time. Policy and management must evolve according to natural dynamics and disturbances as well as social events, economic change, and political values.

Nonetheless, it is clear that ecological sustainability lays a necessary foundation for national forests and grasslands to contribute to the economic and social components of sustainability, making contributions to strong, productive economies and creating opportunities for enduring human communities.

The Committee believes that conserving habitat for native species and the processes of ecological systems remains the surest path to maintaining ecological sustainability. We suggest the use of two general approaches in tandem to conserve these key elements of sustainability.

First, we suggest a scientific assessment of the characteristic composition, structure, and processes of the ecosystems. This assessment should provide an understanding of the "ecological integrity" of the planning area.

Ecosystems with integrity maintain their characteristic species diversity and ecological processes, such as productivity, soil fertility, and rates of biogeochemical cycling. Because ecosystems are dynamic and variable, the concept of the "historic range of variability" is used to characterize the variation and distribution of ecological conditions occurring in the past.

This concept allows one to compare the ecological conditions that will be created under proposed management scenarios to past conditions. The more the prospective conditions differ from the conditions during recent millennia, the greater the expected risk to native species, their habitats, and the long-term stability of ecological processes.

Second, we suggest focusing on the viability of native species themselves. However, monitoring the status of all species and assessing their viability is impossible from a practical standpoint. Thus, it is necessary to focus on a subset of species called "focal species."

The key characteristic of a focal species is that its abundance, distribution, health, and activity over time and space are indicative of the functioning of the larger ecological system.

In monitoring, the habitat needs of the focal species are analyzed, and projections are made of the habitat that will be needed for the species to be considered "viable," having self-sustaining populations well distributed throughout the species range.

Self-sustaining populations, in turn, can be defined as those that have sufficient abundance and diversity to display the array of life-history strategies and forms that will provide for their persistence and adaptability in the planning area over time.

The habitat that will be created under any management scenario is compared to the habitat needed for the viability of each selected focal species. The less adequate the habitat for each species, the greater the risk to native species and ecological processes. Therefore, the Committee suggests a three-pronged strategy:

1. focusing on a set of selected "focal" species and their habitat needs,
2. maintaining conditions necessary for ecological integrity, and
3. monitoring the effectiveness of this approach in conserving native species and ecological processes.

In many cases, national forests and grasslands by themselves are unable to conserve native species and ecological processes. As noted earlier, other landowners and agencies often control key elements of the habitats and ecological systems.

Thus, in some cases, the national forests and grasslands can contribute to, but not ensure, the achievement of ecological sustainability. It is important to note that this approach is similar to the existing regulations implementing the National Forest Management Act.

These 1982 regulations have an extensive section on "Management Requirements" that calls for the provision of adequate habitat to maintain viable populations of existing native and desired nonnative vertebrate species; protection of soils, streams and watersheds; and many other conservation measures.

These requirements were intended to provide a policy framework for sustaining ecological systems within which decisions could be made. In its details of implementation, however, the approach proposed by the Committee for assessing ecological sustainability differs from the existing one, reflecting more than 15 years of experience since those regulations were written.

Conserving habitat for native species remains central to ecological sustainability while broadening the focus from vertebrates to all native species. At the same time, the Committee recognizes that ensuring the viability of all native species, through analysis of individual species, is an impossible task.

To ensure the development of scientifically credible conservation strategies, the Committee recommends a process that includes

1. scientific involvement in the selection of focal species, in the development of measures of species viability and ecological integrity, and in the definition of key elements of conservation strategies;
2. independent scientific review of proposed conservation strategies before plans are published;

3. scientific involvement in designing monitoring protocols and adaptive management; and
4. a national scientific committee to advise the Chief of the Forest Service on scientific issues in assessment and planning.

The Committee recognizes that its role is not to dictate specific management approaches for the Forest Service but to provide advice that the Secretary and Chief may act on as they deem appropriate.

Nonetheless, the Committee recognizes that such concepts as focal species, ecological integrity, and the use of scientific information may involve technical issues and that the Committee thus has an obligation to the Secretary and the Chief to provide some insight on how this framework for ecological sustainability might be converted from concept to application. Therefore, while our approach has not been field-tested, the Committee has drafted the following regulatory language, that, we believe, provides a useful approach to this issue.

- **Wild Horses and Burros Act of 1971** - This act helps in protecting wild free-roaming horses as well as burros from capture, harassment, branding or death. This act states thatit is important to consider them in the area where an integral part of the natural system of the public lands is found presently.
- **The Clean Water Act of 1972** -This act named The Clean Water Act sets the basic structure in order to regulate the discharges of pollutants to waters of the countries like the United States.
- **The Endangered Species Act of 1973** -this acts helps in protecting the species of animal and the plant which are in danger of extinction (that is endangered) currently and the species which can be endangered in the upcoming future (which is basically referred to as threatened). It helps in conserving the ecosystems on which the threatened and endangered species depends such as fish, wildlife, and plants, both through Federal action as well as by encouraging the establishment of state programs.
- **Forest and Range Renewable Resources Planning Act of 1974** -This act named Forest and Range Renewable Resources Planning Act which was found in the year 1974, establishes public land policy as well as guidelines in order to manage, protect, develop and enhance the public lands.
- **Federal Land Policy and Management Act of 1976** -This act which was established in the year 1976, states that the public lands can be managed in such a manner that it will protect the quality

of scientific, historical, scenic, ecological, air, environmental and atmospheric, water resource, and archeological values.

The management can be done wherever required, and eventually, this will preserve as well as protect some specific public lands in their natural condition. This in turn will provide food as well as habitat for fish and wildlife and domestic animals.

Furthermore, this will provide for outdoor recreation and human occupancy as well as its use. This act also states that the countries such as United States shall receive fair market value for using the public lands and their resources until unless it is otherwise provided for by law.

- **Public Rangelands Improvement Act of 1978** -This act came into existence in the year 1978 and this establishes and reaffirms the national policy as well as the commitment to inventory and furthermore, helps in identifying the current public rangeland conditions and trends. It helps in managing, maintaining and improving the condition of public rangelands so that, they become much productive and feasible for all rangeland values considering the management objectives and the planning process for land use. This act charges a fee so as to use public grazing which is equitable; it also, continue the policy that is for the protection of wild free-roaming horses and burros from capture, branding, harassment, or death. At the same time, it facilitates the removal and disposal of excess wild free-roaming horses as well as burros which pose a threat to themselves and their habitat and also, to other rangeland values.
- **The Rescission Act of 1995 (Public Law 104-19) -** This act came into existence in the year 1995 and this required each National Forest to establish as well as adhere to a schedule for completing the NEPA analysis and decisions that are for all grazing allotments within a period of 15 years.

5.5.1. Regulations

There are a number of examples for the policy which relates to the range resources as well as coordination of the range activities of the agencies such as USDA and other executive agencies, organization and individuals and some of the examples are mentioned below:

- Departmental Regulation, Number 9500-5 - dated December 15, 1983; Subject: Policy on Range.

- Executive Order 12548 - Grazing fees, 1986 - provide for the establishment of appropriate fees for the grazing of domestic livestock on public rangelands.
- Secretary's Administrative Order of August 1963, Administration of Lands Under Title III of the Bankhead-Jones Farm Tenant Act; Establishment of National Grasslands.

5.5.2. Policies

Some of the examples of the policies have been mentioned below which are followed by some of the countries such as the US:

- National Grasslands, A Primer (1997) - This is a document that helps in identifying and interpreting the laws as well as the regulations which are applicable to the administration of the National Grasslands.
- Forest Service Rangeland Management Directives - These directives collectively make up the Forest Service's Rangeland Management Manuals and Handbooks.

5.6. PERMANENT PASTURE

The grasses can be grown naturally (that means self-seeded) or through cultivation (that is sown). The land is used to grow grasses or the herbaceous forage, which has not been included in the crop rotation for five years or more.

Figure 5.3. Pasture Cow Agriculture.

Source: Image by Pixabay

Now the definition of permanent pasture has changed.

- The definition of permanent pasture, the application of which was done before the reform of the **Common Agricultural**

Policy (CAP), excluded areas that are considered under the **Single Payment Scheme** or some **agri-environment scheme** options.

- Considering the reformed CAP, it can be said that the permanent grassland and permanent pasture are together referred to as permanent grassland. Permanent grassland can include other species, like the shrubs and trees which can be grazed, provided that the grasses and other herbaceous forage remain predominant.

The definition of permanent pasture is said to be important, and the reason is that according to the **European Union** legislation, each **member state** is required to maintain the ratio of land under permanent pasture in relation to the total agricultural area. If there is a decrease in the area of permanent pasture below a certain percentage, then it is important for a member state to take steps so as to prevent further reduction. Farmer converted permanent pasture into other uses and such farmers might be obliged to reconvert land.

Case Study: Grassland Tenure in China: An Economic Analysis

The government has initiated the Grassland Policy where grassland use rights were distributed to groups of households and cutting lands were distributed to individual households. The distribution of grasslands and cutting lands was done on the basis of livestock numbers and thus, implicitly, household population and labour force. The legal and regulatory framework has evolved since, with the passing of the national Grassland Law in 1985, the Xinjiang Grassland Law in 1989, and various regional government grassland regulations.

The group tenure and creation of fuzzy boundaries in grasslands facilitated the abatement of environmental risk by allowing flexibility and mobility. First, in terms of the characteristics of tenure, grasslands and cropland share in common collective ownership and de jure household tenure but little else. The de facto unit of tenure in grasslands is groups, implying joint use of resources, whereas in croplands individual household tenure prevails.

Grassland use rights have a longer term of duration (50 versus 30 years) and grassland allocation has been very stable. In cropland areas, in contrast, periodic administrative reallocations of land have been relatively common despite formal policy stipulating first a 15 and then a 30-year term of use rights.

Another point of contrast concerns the monitoring and enforcement of fuzzy boundaries in grasslands whereas in croplands they are precisely monitored and enforced. Finally, pastoral households use larger parcels of land, dispersed over a much wider area, than their cropping counterparts.

5.7. FAO LAND COVER CLASSIFICATION SYSTEM

The Land Cover Classification System (LCCS) is considered to be comprehensive as well as standardized a prior classification system. There are two main phases in the classification which include:

i An initial Dichotomous Phase, in this phase eight major land cover types are distinguished, and

ii A subsequent Modular-Hierarchical Phase, in which the set of classifiers and their hierarchical arrangement are tailored to the major land cover type.

By adding the environmental attributes (such as climate, altitude, landform, soil, lithology and erosion), further definition of the Land Cover Class can be achieved. Specific Technical Attributes are related to particular technical disciplines (such as floristic aspects, crop type or soil type). Grasslands come under "Cultivated and managed Terrestrial Areas" and "Natural and Semi - Natural Terrestrial vegetation".

Figure 5.4. Landforms

Source: Image by Wikimedia Commons.

There are different relevant definitions in relation to the grasslands in the classification system and these include:

- Herbaceous: These are the plants which do not have persistent stem or shoots above ground and furthermore, it lacks definite firm structure. These are referred to as herbaceous. There are two categories, which depends upon the physiognomy such as Graminoids and Forbs.
- Graminoids: The graminoids include all the herbaceous grasses and other grass- like plants which have narrow leave. According to the taxonomic definitions, these are not grass.
- Guidelines: The vegetation named graminoid can be defined by the presence of more than 75 percent of Graminoids in the herbaceous coverage. They do not have an upper limit of height:

the only condition that exists is the physiognomy of the plant.

- Cover: The cover can be defined as the proportion of a specified area of the ground, substrate or water surface which is covered by a layer of plants. The cover is considered at the greatest horizontal perimeter level of each and every plant in the layer. There can be a distinction which is made between closed (more than 60 – 70 percent), open (70 – 60 percent to 20 – 10 percent), closed to open (between 100 and 15 percent) and sparse (20 – 10 percent to 1 percent). The herbaceous plants are seasonal in character; therefore, it is important to note that the cover of herbaceous vegetation is always considered at the time of its fullest development.
- Height: Considering a certain layer, the height is measured from the ground to the average top of the life form which is going to be assessed. Height sub-divisions are greater than 30 down to 3 m for Trees; 5 m to 0.3 m for Shrubs; and 3 m to 0.03 m for Herbaceous.

5.8. CONCLUSION

The policies and regulations are the ones that can affect the grassland in a positive as well as in a negative way. The policies can further be classified into two broad categories. One of them is the government policies that cover the comprehensive landscape issues such as Grassland Household Contract System (GHCS) in China or the Federal Land Policy Management Act (FLPMA) in the USA. The other one is narrow in scope policies which most of the time provide subsidies in order to accomplish the particular production or environmental objectives.

Comprehensive policies are the one which is developed in order to address the major country -wide grassland issues most of the time. But after some time, it can result in the unintended negative outcomes. The narrow subsidy - based policies accomplish particular goals, but it tends to come at a monetary expense to society.

The conclusion that can be drawn from the chapter is that any policy which affects the grasslands is required to be carefully monitored over time. It is important for the government that they should discontinue or dramatically change the policies which have become ineffective. The government should enhance the policies which are providing useful production, environmental and societal benefits.

REFERENCES

1. CASE STUDY SUMMARIES. (n.d.). [ebook] p.2. Available at: https://www.iucn.org/sites/dev/files/content/documents/casestudy_summaries.pdf [Accessed 28 Feb. 2020].
2. Cheung, D. (2019). *Law vs Policy. What's the Difference?* [online] Career & Internship Center | University of Washington. Available at: https://careers.uw.edu/blog/2019/10/13/law-vs-policy-whats-the-difference/ [Accessed 28 Feb. 2020].
3. Fs.fed.us. (n.d.). *Rangeland Management Laws, Regulations, and Policies.* [online] Available at: https://www.fs.fed.us/rangeland-management/aboutus/lawsregs.shtml [Accessed 28 Feb. 2020].
4. Implementing the Laws and Policies Governing the National Forests and Grasslands in the Context of Sustainability. (n.d.). [ebook] p.1. Available at: https://www.fs.fed.us/emc/nfma/includes/cosreport/cos-ch67ab.pdf [Accessed 28 Feb. 2020].
5. National Range Resource Management Policy. (2014). [ebook] p.4. Available at: http://extwprlegs1.fao.org/docs/pdf/les149694.pdf [Accessed 28 Feb. 2020].
6. S, P. (n.d.). *Difference Between Law and Policy | Difference Between.* [online] Differencebetween.net. Available at: http://www.differencebetween.net/miscellaneous/politics/difference-between-law-and-policy/ [Accessed 28 Feb. 2020].
7. Uk.practicallaw.thomsonreuters.com. (n.d.). *Permanent Pasture.* [online] Available at: https://uk.practicallaw.thomsonreuters.com/3-518-9434?transitionType=Default&contextData=(sc.Default)&firstPage=true&bhcp=1 [Accessed 28 Feb. 2020].
8. Swedish Board Ag., 2015
9. MAFF, 2015
10. MOE, 2015

6

Improvement and Rehabilitation of Grassland

CONTENTS

The chapter of improvement and rehabilitation of grassland talks about the basic significance and requirement of the improvement and rehabilitation of the grassland. This chapter also talks about the state of grassland. This chapter also includes the development of grassland, improvement of the grassland, and rehabilitation of the grassland which includes the use of water, natural salt-licks, and trees and shrubs.

This chapter also explains how to enhance the quality of the grassland such as the quality of swards, quality of the nutritional, and health of the animals. This chapter also explains the effect of excess nitrogen that as an outcome decreases the palatability of silage, which is consisting of phosphorus deficiency reduces digestibility, potassium regulates the control rate of leaf growth, magnesium deficiency can produce animal health issues, sodium improves palatability, as well as some other crop management practices that impact the quality of the grassland.

This chapter also addresses the grassland renovation and grassland improvement. This chapter also provides highlights on the rehabilitation of the grassland with the help of clearing, bush control, fire, and fencing. This chapter also explains the herd management such as stocking rates and the stock distribution, and scarification. This chapter emphasizes the social, economic and environmental factors of grassland improvements such as tenure, markets and trade, herder organization and community participation, diversification, and grassland in the environment.

6.1. INTRODUCTION

The north equator and the south equator are encircled with the world's grassland. The distribution of the grassland is identified with the help of the climate, topography, and soil. The existence of them is because of the principally grazing animals, biotic factors, and fire.

Also, man had extended the grassland of all across the globe by clearance of the forests for cropping and for grazing his or her domestic animals. Certainly, the cultivated grasslands that are established in the areas whose natural climax is forest are the most intensively used and, with respect to the agriculture they are most productive and effective.

The evolution and the ecology of the grasslands are very complicated and controversial (Moore, 1964). The activities of humans have so modified the grassland all across the world, that, in the present interval of time, it is sometimes very difficult to differentiate the natural from the man-made.

The casual observer viewing the brown top clad foothills of New Zealand's South Island can be quite surprised in order to learn that the species (which is Agrostis tenuis) only arrived in New Zealand with the European settler. There is a disagreement as to whether the natural grasslands show a climax society and about the degree to which the allocation of the grasslands is identified with the help of climate against the factors such as soil and biotic factors (Moore, 1964).

The constrained rainfall, or at least the frequency of a dry season, more particularly when reinforced with the help of other factors, especially fire and grazing, gives the idea to be a common factor that prevents the formation of a forest canopy.

6.2. THE STATE OF THE GRASSLANDS

In the present interval of time, the condition of the grassland all across the globe is very distinct, but in several numbers of cases, the condition is far from the satisfactory level. Long-term historical data on pasture is scarce in most of the areas. In this way, the level of shift or change or degradation has to be inferred from the existing condition.

In all but the coldest as well as driest regions, the large areas of the better land have been cleared for crops, leaving the poorer pasture to the extensive stock rearing. In conventional areas of subsistence farming, this is due to increment in the population of the human among agricultural groups; elsewhere expansion of the crops into the marginal lands has been in the hope of the profit, which has every so often not been realized.

Most of the grassland of all across the globe are on poor quality land. According to Buringh and Dudal (1987), only about 16.66 percent of the grassland all across the globe are on high and medium category land. On the other hand, the rest 83 percent are on low to zero classes. In this way, the potential for further clearing of the grassland for sustainable cropping seems quite low.

Cultivation of the grassland has led to the issues of access to the water for stock and wildlife, obstruction of migration routes and the fragmentation of the wildlife habitat, and loss of lean season grazing. Increment in the population pressure and poverty, more particularly in the savannah zones of Africa, has driven subsistence cultivators further into the drylands.

Figure 6.1. The state of the grassland.

Source: Image by PxHere

Typically, it is the best soils and regions along watercourses and other sources of the water that are developed first, these have generally been lean-season or emergency grazing lands of the pastoral groups and their clearing can upset the grazing systems, and it takes into the direction towards the degradation of the remaining grassland.

Crop production is not the sole invasive utilization of the grasslands. Forestation can constraint the migration routes as can fencing for the game exclusion (or safeguarding). Declaration of the game reserves in pastoral regions can impact the herding, and mining and extraction of oil also produce harm and damage.

In several regions of eastern Africa, pastoral systems are contracting with the help of expansion of cropping, and the grassland is more and more being combined into the farming systems. National land tenure legislation is not linked to the conventional grazing rights and puts pastoralism at a disadvantage as compared with the crops. The pastoral vegetation, never the less, is resilient and recovers well after a drought.

In South Africa also, much of the better grassland in commercial regions has been cleared for annual cropping, and in communal regions the better watered land has been transformed into a patchwork of crops and thicket.

The grassland vegetation is typically resilient, never the less, there is some degradation in the driest regions. Bush encroachment is an issue in several numbers of vegetation types. Low returns from a broad commercial stock rearing in the dry areas are taking into the direction to depopulation. On the other hand, in some of the cases, the enterprises are shifting to the game ranching.

The population of humans in West Africa has been elevated enormously so the grassland has declined and near about all the cultivated grassland in the better -watered regions. In addition to this, vast tracts of semi-arid marginal land, are now under subsistence crops. The authority of the tribes, which had regulated the grazing practices, broke down in several numbers of nations once they gained acquired their independence, leaving the grassland as an open-access resource.

In several regions of North Africa, the population of the human has been elevated with higher pace and conventional authorities and grazing rights have broken down. Several semi-arid lands have been cultivated for the unsustainable annual cropping. Livestock numbers have increased and more can be carried with the help of lean season with the use of purchased- at one time subsidized, which is cereal and concentrates.Transhumance cycles are mostly greatly cut or herds have become sedentary. The uprooting of shrubs for fuel produces severe damage. All the grazing land is overstocked and degraded, every so often seriously so.

The conditions in Patagonia contrasts sharply with that of East and South Africa. In a little over the time interval of a hundred years from the introduction of sheep ranching, the vegetation has been severely modified with the overgrazing, primarily in the last five decades. Ranching is on private land with the last paddocks and little grazing control within them. Guidelines for pasture management were only established in the year of 1980s and have yet to make a strong influence. The Campos is comparatively well- watered, and all stock rearing is commercial. Pasture technology is well developed. Introduced forages are used to palliate seasonal differentiations in fodder availability and quality, either as sown pasture or by over-seeding. The value of suitably managed native pasture is more and more appreciated. Grasslands are commonly in good condition.The well- drained parts of the Pampas are now farmland, where field crops are every so often in rotation with sown pasture. The flooding pampa is still exploited as grazing land, creating stock for fattening elsewhere.Several numbers of the pastures of the Gran Chaco deteriorated critically through the course of the twentieth century. Before the time period of 1920, it had been almost unmanaged extensive grazing. The provision of watering points permitted far more of the vegetation to be utilized, leaving little in order to carry fire.

In this way, together with the unregulated felling of the forest takes into the direction to the invasion with the help of undesirable thorny vegetation. Bush violation carried with the help of overstocking and deficiency of the grazing management is very critical, which takes into the direction towards the erosion, reduction of livestock production, and loss of wildlife habitat.

The economics of the herbicides and the mechanical clearing is not clear. In most of the sub-tropical areas of the extensive grazing, the strategic use of the pasture resting and the regulated fire is the only economic approach of keeping bush in check.

In the grasslands of Central North America, the better-watered tall-grass prairie is now mostly under crops. Vast regions of drier land have also been cleared for cropping purposes, several of them were marginal for crop production and suffers from the periodic drought, which takes into the direction to the "dust bowl" of the time period of the 1930s.

Several numbers of marginal cropland ahs been reseeded or transformed into forested, with considerable government support. Low returns from farming and the isolation of several numbers of rural societies are resulting in urban drift. The modern trend is led towards the larger-scale landscape management, now that information collection and manipulation technology are available.

6.3. GRASSLAND DEVELOPMENT, IMPROVEMENT AND REHABILITATION

Most of the grassland, whether commercially or conventionally managed, has needed some development inputs to make stock-rearing possible or more effective. All grazing resources have to be taken into the consideration and these cover much more as compared to the herbaceous stratum.

Figure 6.2. Grassland development, improvement, and rehabilitation.

Source: Image by PxHere

Types of Grassland Resources

6.3.1. Water

Water is the prime determining factor in stock management in most broad grazing lands. In the regions that are dependent on seasonal surface water, the stock must move out once sources have dried. Enhancement and development of water supply with the help of developing water points or improving existing ones, and clearing of undesirable vegetation in order to permit free access for stock and better grass growth, are very common to both of the systems, and the provision of materials or conventional salt licks is frequent.

The availability of water is a factor in identifying several numbers of migration patterns in mobile systems. In both East and West Africa, the conventional rules govern pastoral water use, and in very dry regions is a more important resource as compared to the grazing.

In regions with the very cold winters, according to the study of Mongolia, the surface water freezes; wells provide water, but, in their absence, herders may have to abstract the water from the below ice, melt snow or have stock eat snow in order to find the water in severe winter weather conditions, dehydration may be as damaging to stock as lack of food.

Without the water development, the stock would be constrained to the regions that are close to the permanent source of the water all across the dry season, and large regions of the grassland that would not be useable for the livestock production. From ancient times, stock watering points have been established in order to make sure the year-round supply of water within the grazing region of a group or in order to make grazing land accessible.

Access to the water is mentioned as a limiting factor in order to utilize some grazing regions in South Africa. According to the studies of South America, the creation of water supplies has made stock rearing possible in large regions of Patagonia and the Gran Chaco, and the significance of the water development is stressed in Australia.

On the other hand, the breakdown of the mechanized wells after de-collectivization has rendered large tracts of Mongolia is inaccessible with respect to the domestic livestock. The kind of livestock that impacts the frequency of the water, camels are by far the most hardy. On the other hand, most of the cattle and stock can only graze for about two days away from the water in hot, dry climatic conditions.

6.3.2. Natural Salt-Licks

Natural salt-lick (that are the deposits or the salt springs that are licked by animals) are valued in several numbers of zones of extensive grassland and also, they are used by both livestock as well as grazing wildlife. Herders in, for instance, sub-Saharan Africa and Mongolia take their herd long distances for periodic access to the licks.

Salt is, of course, generally given to the livestock, either alone or in proprietary blocks that can be consisting of other minerals as well. On the other hand, the herbage is lacking with minerals, which are very important with respect to animal growth, their well-being ,as well as productivity, suffers.

Figure 6.3. Natural salt-licks: grassland development.

Source: Image by wikimedia commons

The absence of the phosphorus is widespread and more particularly, is acute in regions of sub-Saharan Africa. Various numbers of minor components may be lacking in the localized regions, the pastures of the Kenya Rift Valley between the Nakuru and Naivasha were tarnished for poor livestock performance. This "Nakuritis" was identified as the lack of cobalt and has since been discovered on several numbers of other grasslands.

6.3.3. Trees and Shrubs

Trees and shrubs are the most significant and necessary aspects with respect to the several numbers of the grasslands, more particularly of savannahs. Some of them are very significant and useful as well, on the other hand, others are invasive weeds. Trees deliver valuable shade through the course of hot climates and seasons and trees give shelter in winter as well. Some trees are browsed and may be lopped for fodder - their fruits can also offer a valuable feed.

An extensive range of genera taking participation and their management is still poorly acknowledged. The tolerance of most of the woody species to stock-feeding regimes and lopping still needs to be considered. Never the less, a good deal of work has been done on Leucaena.

Woody vegetation offers branches as well as poles for the purpose of building and making corrals as well as firewood. With respect to this, firewood is scarce, extreme and unwanted cutting of the trees causes serious harm to the environment, as in steppic conditions, where much harm and damage is produced because of the uprooting sub-shrubs for fuel.

Some of the trees also offer fruit which is valued with the help of local people. These kinds of trees may be reserved selectively and given some guard and safety. Woody vegetation is, never the less, every so often invasive, more particularly in tropical regions and sub-tropical conditions. The violation of bush is commonly considered as a sign of poor management and overgrazing.

6.4. HOW TO IMPROVE GRASSLAND QUALITY?

The three aspects to grassland and pasture quality are nutritional quality, sward quality, as well as animal health. Each of these aspects is greatly impacted and enhanced with the help of a balanced nutrition programme.

Figure 6.4. The quality of the grassland.

Source: Image by PxHere

Nutrition impacts all of the aspects of grass quality. There are three characteristics to the grassland and pasture quality: sward quality, animal health, and nutritional quality.

6.4.1. Sward Quality

Sward quality is impacted more as compared to the other agronomic practices as well as grazing management, never the less, maintaining a good level of soil fertility and incorporating a balanced crop nutrition program should be a significant part of grassland management.

There is also indirect proof that several nutrients which are consisting of molybdenum, zinc, potassium, boron, and manganese that decrease the infection of disease in cereals. In this way, with the help of extension, this is also likely to be the same for forage grass species.

6.4.2. Nutritional Quality

Nutritional quality of the grass is generally evaluated with the help of digestibility, the content of protein, and content of dry matter which can all be impacted with the help of crop nutrition.

Digestibility relies on the growth stage and nutrition, that is young, leafy swards have a higher level of D and a higher level of ME as compared to those that have started to head or contain a lot of dead material. Grasslands with a balanced nutrition strategy will be more palatable as compared to those that are under fertilized.

The levels of proteins rely on the grass growth stage and are impacted with the help of crop nutrition, more particularly in nitrogen fertilization. Levels elevated in the initial time of springs. The formation of protein relies on the potential of the plants in order to take up the content of nitrogen from the soil. In this way, they can be impacted with the help of the application of the nitrogen but also with the help of the content of sulfur, pH of the soil, and content of potash.

The content of the dry matter is the most variable and the least controllable. Low dry matter is every so often linked with the wet weather and poor growing conditions. As well as having a direct impact on the production, this also tends to decrease the quality and take into the direction lower intake on a daily basis. In addition to this, in silage, a high dry matter content (which is more than forty percent) can lead to a lower intake.

6.4.3. Animal Health

The animal health and trace element nutrition should also be well thought out as a characteristic of the grass quantity. Grazing animals have several numbers of different needs for trace elements with respect to those that are needed for the growth of the grass. In this way, it is very significant to make sure that these elements are present in the grass in the appropriate or in adequate level in order to fulfill the nutritional requirement of the animals.

Selenium and sodium both are very important with respect to the health of the animals. But, in the context of plants, they are not required, along with that, guidelines for the other kinds of trace elements, for example, magnesium, copper, and zinc are higher in the grazing animals as compared to the present in grass.

6.5. NITROGEN EXCESS REDUCES PALATABILITY OF SILAGE

High rates of nitrogen can also produce issues with respect to the fermentation, and this is due to the presence of excess nitrate that has a negative effect on the fermentation process and will produce silage that is less palatable and the animals will be less keen to eat it.

Commonly, Nitrate is taken up by grasses much faster as compared to when it is incorporated into proteins and until applied this excess known as luxury uptake is stored in the leaves. Extra content of nitrate will be present if inadequate time is allowed between the application and mowing and can also take place under the conditions of poor growth, for example, cool or low temperatures, and low light levels.It is also an issue if there is a dry spell after the application, when nitrate cannot be taken up by the roots, followed by a course of wet weather that, as an outcome, in luxury uptake. The plant cannot transform it into protein with a quick rate, in this way it accumulates in the plant.

High rates of nitrogen can produce a decrement or decline in crop sugars as they are utilized in order to provide energy for the elevated rate of plant growth and for the development of plant proteins. In addition to this, this elevated growth rate can as an outcome, lead to the lower crop dry matter content, never the less, in practice this is every so often is not meaningful.

An extra amount of nitrate can be avoided with the help of the Grassland Rules that are mentioned below in order to apply nitrogen at the rate of no more as compared to 2.5 kg N/ha/day.

6.5.1. Phosphorus Deficiency Reduces Digestibility

Phosphorus plays an important and vital role in several numbers of plant metabolic processes as well as in enzyme activities. In this way, even though the demand for the phosphorus is low as compared to that of nitrogen its availability is very important. Where there is a lack or low availability of the phosphorus, the digestibility and the nutritional quality of the forage will be decreased.

Figure 6.5. A phosphorus deficiency in corn grown in the coastal region.

Source: Image by Flickr

Phosphorus is very immobile in soil and its availability is limited by both pH as well as by distance from plant roots.

6.5.2. Potassium Controls Rate of Leaf Growth

Potassium is the nutrient taken up in the greatest quantity with the help of grassland swards.

Potassium plays an extensive role in the context of a plant to impact the nutrient uptake, photosynthesis, rate of growth as well as feed value.

If proper amounts of potassium are not available, the rate of production as well as rate of growth will be limited. Never the less, there can be dangers with respect to the health of an animal if excessive amount of potassium is applied with an increased risk of hypomagnesaemia.

Figure 6.6. The deficiency of potassium in plants.

Source: Image by wikimedia

6.5.3. Magnesium Deficiency Can Cause Animal Health Issues

Magnesium also requires to be well thought out. Magnesium is a very important nutrient, and there are several numbers of situations where magnesium is lacking. The lack of magnesium will as an outcome, produces a considerable amount of decrement in the quality of the grassland and can have critical results on the health of the animal.

Sodium ,as well as selenium, are both important and significant with respect to the health of the animal but they are not needed by plants and guidelines for other trace elements such as magnesium, zinc and copper are higher in grazing animal comparative to the present in grass.

6.5.4. Sodium Improves Palatability

Sodium is very vital in the context of animal nutrition and every so often the grass is consisting of an inadequate amount in order to fulfill the necessities of the grazing animals.

Symptoms of deficiency of sodium in cattle may be consisting of low appetite, decrement in the fertility and decrement in the production of the milk. Sodium also assists to increase the palatability of forage, in this way elevating the dry matter intake.

Fertilizers that are full of sodium, elevates the level of the sodium in the grass and have also been shown to elevate the effectiveness and productivity.

6.5.5. Other Crop Management Practices Influencing Grassland Quality

Growers can also impact the quality of the grassland with the agronomic practices that are mentioned below:

1. Maintaining the correct amount or content of pH in the soil.
2. Maintaining good soil structure and minimizing the soil compaction.
3. Enhancing and maintaining the drainage of the soil.
4. Maintaining a high proportion of the desired grass species within the sward in order to maximize the nutritional value.
5. Daily basis- reseeding and/or oversowing to maintain high -quality swards across the farm and selecting varieties for quality as well as yield.
6. Topping to remove the course material and harrowing in order to remove the dead matter and open the sward, both of which will encourage tillering.
7. Managing grazing with efficacy in order to prevent the damage and minimize rejected regions.

6.6. GRASSLAND RENOVATION AND GRASSLAND IMPROVEMENT

Permanent grassland is a potential, but they are not still not considered as an optimal used resource with respect to the production of protein and meets the new political targets for farmyard protein production without soja perfectly. Chances for elevating the production of the protein from grassland are in general early cutting, the increment of the nitrogen fertilization and an increment in the N-efficiency and enhancement of the botanical composition with the help of higher percent of legumes.

More particularly, the swards of permanent grassland in South Germany have every so often high percent of herbs as well as weeds with only slight reactions to the fertilized nitrogen (Elsaesser, 2005). On the other hand, grasses respond more to the fertilization of nitrogen.

Grassland renovation in order to enhance the amount of grasses can consequently be a possibility to elevate the productivity of the grassland. Never the less, as an alternative to the anticipation of elevated DM production in the years following the treatment (Taube and Conjin. 2007), reseeding of permanent swards might not always outcomes in net profits (Soegaard et al. 2007, p. 97).

In addition to this, the total costs of renovation in the context of lack of production in the first year after treatment were - so far - not calculated exactly. With respect to the evaluation of the success of the grassland renovation method, the enhancement of the botanical composition has to be taken into consideration. Moreover, the numbers of methods and approaches that are used in agricultural practice are to be examined.

6.6.1. Over and Under Grazing Are Mistakes in The Pasture Management

– **Overgrazing**: Higher demand on feed then the grassland offer.

Effect: too long grazing period on the same grassland area. Too short grazed grasses with too low resistance.

– **Under grazing**: Feed offer is much higher as compared to the demand for grazing animals.

Effects: Partially too old grasses with only low input and high grazing residuals.

An assortment of horses helps to protect the single plants. the grassland sward switches to an increase of not grazed plants which have the chances in order to have generative reproduction. As n outcome, the botanical composition of the grassland alters.

From this high plant such as Urtica dioica, thistles, docks, or grasses such as *Deschampsia caespitosa* have advantages. But also, more valuable plants such as *Dactylis glomerata* or *Festuca arundinacea* are not or only less eaten.

In addition to this, another impact also occurs. Already deep grazed regions will be more and more short grazed. Because of an imbalance between the grazing region and the number of horses grazing on it, the tic period of grazing is very much longer in some of the regions and the phase of rotation is very less.

In this way, the essential recreation coursed for important plants were reduced as well and the grass swards have not sufficient reserves for a good and rapid regrowth. Only good grassland management can provide assistance in order to avoid these kinds of impacts or effects.

6.6.2. A Good Grassland Management Has the Obligation to Understand How Management Faults Work

Overgrazing can regulate itself even if the actual grazing region is restricted. With the help of increment in the rank patches or with the help of aging of grassland growth. The animals bite frequently the regenerating and new plants to create bald patches in which annual herbs or grasses can immigrate (for example, *Capsella bursa pastoris*, *Poa annua*, *Camomilla* spec. and *Stellaria media*).

In addition to this, an increment of foot traffic ,as well as bite-resistant species (plants with thorns, dense hair or with toxins, for example Ranunculus or *Holcus lanatus* species), and not detectable plants in rosettes (which are daisy or common plantain) will take place. Simultaneously, existing weeds (for example, thistles, and *Rumex obtusifolius*) will find an ideal setting and spread even more closely.

Under grazing take place for example, if the start of grazing is too late and is too old and too far developed swards or if there are too little horses per area. This takes place when the horses have sufficient or even too much space for their movement.

6.7. REHABILITATION OF GRASSLAND

Most of the grasslands, whether it is commercially managed or it is conventionally managed, have needed some kinds of development inputs in order to make stock-rearing possible or more productive. All of the grazing resources have to be taken into consideration and these cover much more as compared to the herbaceous stratum.

Pasture Development Methods

6.7.1. Clearing

After the water supply, clearing is a communal part of emerging wide-ranging grassland for grazing. Where land is being technologically advanced with respect to the crops or sown pasture, clearing could be consisting of some removal of stones, termite hills as well as other obstructions, but, for wide-ranging grazing, clearing generally consisting of removing or thinning woody vegetation in order to enhance the access as well as the growth of the grass or to decrease the tsetse fly habitat.

With respect to the conventional systems, fire is the commonest agent for clearing or controlling trees along with the shrubs. Particular types of equipment are being into use concerning the large-scale commercial clearing, such as drag chain, root plows, tree-pushers, bulldozers, and root rakes

On the other hand, for shrubs, with the help of several kinds of rollers and shredders; the debris may be burnt. The level of clearing or thinning will rely on the original vegetation and the utilization to which it is put, but it is commonly partial as well as selective, leaving significant and meaning trees, shelter and shades.

Figure 6.7. Rehabilitation of grassland.

Source: Image by PxHere

Strategic thinning of woody vegetation plays a very important role in pasture development and enhancement, but it must be done within the context of the ecosystem involved, from the time, the removal of trees and their replacement with the help of crops and annual pastures can make major alterations and modification to the hydrological cycle and can lead to critical salination of soils.

6.7.2. Bush Control

Bush control is a very important aspect in several numbers of grassland types; it is a maintenance activity, on the other hand, bush clearing is development.

Violation of bush generally depicts defects in the management system and it is linked with high grazing pressures; various kinds of mechanisms are taking participation according to the vegetation type as well as the management system.

Unpalatable shrubs can provide assistance in the increment when the more palatable ones they are overgrazed; if little dry herbage remains in the non-growing season there may not be hot enough fires to control the bush.

6.7.3. Fire

Regulated fire is one of the major factors in order to find out the arrangement of grasslands and a widespread and powerful tool in grassland management. The impact of the fire relies on its seasonality, intensity, frequency and type. The intensity of the fire relies on the kind, abundance of fuel, and structure.

Fire is used in order to remove the unpalatable grass and allow regrowth to the plants and access to the young herbage with the help of grazing stock. It every so often stimulates regrowth and supplies a green bite when most required. Fire is also used, as discussed above in the chapter, in order to regulate the woody vegetation.

6.7.4. Fencing

Fencing is extensively used in the development of commercial grazing enterprises in order to delimit properties and subdivide them for ease of management. Block size is usually large on low-yielding grasslands from the time, the fencing ,as well as fence maintenance, are very expensive in nature; this can take into the direction towards the uneven stock allocation. Fences are also used in order to safeguard the forages and hay land within properties.

6.7.5. Grassland Improvement

"Improvement" of wide-ranging natural grassland with the help of introduction of specific kind of local or exotic grasses and legumes has been done experimentally in most of the better- watered zones, and is utilized with the help of some commercial systems. Indeed, it is, along with sown pasture, common in commercial mixed farming and more intensively regulated or organized grassland.

Generally, the techniques are consisting of at least temporary suppression of the existing vegetation (with the help of fire, hard grazing, herbicides or mechanically, alone or in combination) and variation in levels of disturbance of the soil surface; fertilizer is every so often is used, and when legumes are introduced to a region for the first time inoculation of their seed with the appropriate Rhizobium is a wise precaution.

It has to be noted that the choice of species and cultivar to suit climate conditions, soil and ultimate use is very vital. On the other hand, there is a very wide range of genetic material of pasture crops that are available. It might be difficult to match them to the new areas.

Also, it is difficult to find out the commercial quantities of seed of locally adapted cultivars and ecotypes. Often, care in management is required in order to guarantee the longevity of the introduced species, and maintenance fertilizer may also be needed. The success of, and in part the need for, over-seeding is reliant not just on the climatic conditions and soil but it also depends on the vigor and aggressiveness of the native vegetation.

6.7.6. Grassland Rehabilitation

Degraded grassland is considered as a symptom of weakness in the pastoral production system. These weaknesses have to be recognized and these weaknesses should be dealt with before taking any further action. Where rehabilitation of grassland is needed it should be through the management methods. It should be with or without water-spreading.

It has been observed that overseeding degraded rangeland is rarely an attractive option and the dependence has to be put on making the most of the recovery of the natural vegetation. Guaranteeing even grazing over an area, keeping all the stock numbers within reasonable limits, and evading localized overgrazing can be helpful.

The process of degradation of pasture can have impacts more serious as compared to the reduction of available grazing; increased runoff can give rise to flooding and siltation of more valuable land and infrastructure lower in the catchment

6.8. HERD MANAGEMENT: REHABILITATION OF GRASSLAND

In commercial systems, the aim of the management is to improve the animal status and generally focuses on one species, or at the most two, species. Common management practices to that end consist of the division of herds into some categories so that they get the correct treatment, evading underage and unseasonable breeding; controlling parasites and predators; providing appropriate veterinary care; and using as well as maintaining breeds that best suit the land and potential markets.

Often, the commercial properties are ring-fenced and it is divided into paddocks. This allows herd division and, in some of the cases, rotational grazing or resting of part of the grassland. The choice of species and breed is in part is determined by the pasture and the climatic conditions. In commercial systems, market needs are vital and, on extensive grazing, beef cattle and sheep predominate.

Figure 6.8. Herd management: rehabilitation of grassland.

Source: Image by Flickr

It has been observed that in warmer climates, Often, zebus or cattle with some zebu blood, developed locally but they are becomingly increasingly popular. Generally, cattle are specialized breeds. These breeds are not dual or multipurpose. Sheep breeds are usually raised in Australasia, it dominates the other commercial areas studied.

Traditional systems, while selling the livestock as well as livestock products, are designed mainly to provide subsistence and security to the herders. Generally, livestock is multipurpose, producing milk, fiber, meat, transport, hides, draught and manure. This is also used in treeless lands as fuel.

It has been observed that they often keep several species. These species may be herded separately; this helps in providing a wider variety of products.

6.8.1. Stocking Rates and Stock Distribution

The regulation of the stocking rate and the management of the spatial and temporal distribution of livestock are considered as the basis of grazing management. The amount of livestock that a specific area of grassland can carry is not just reliant on its botanical composition since it has to take into consideration the management objectives of the Graziers and the availability as well as siting of other grassland resources, particularly water.

It has been observed that the extensive grasslands are not homogeneous but generally, there is show spatial heterogeneity according to moisture as well as the fertility gradients. It has to be noted that stock may tend to concentrate on the better grassland and it overlooks the poorer sites or those sites that are farther from water. Some pastures or grasslands may be suited to grazing at specific seasons or, as in alpine grasslands, it is only available seasonally.

Stocking must be seen in the context of the entire area that is available and management decisions that are made in the light of local knowledge, be it the farmer or landowner who has proper knowledge about his property or the herding group which is a traditional knowledge of their grazing grounds.

It has to be noted that extensive grazing is managed at the landscape rather than at the local scale. Since the productivity of natural pasture, particularly in drier climates, differs widely from year to year. In addition to it, the maximum amount of livestock that can be raised thereon also differs.

6.8.2. Stratification

In commercial systems, the process of stratification of livestock production, usually the fattening stock under more favourable conditions than those conditions in which they were raised – is widespread. It can be a means of lessening the numbers that are carried on pastures through the lean season.

It also helps in speeding up the production cycle since, if stock is shifted to better and a proper pastures and feedlots, they avoid the growth checks and weight loss that is associated with extensive grasslands because of the scarce, low-quality fodder at the time of the lean season. There is a need for meat form quickly grown stock for the markets.

Usually, stratified production systems are linked with the commercial livestock sector. It has been observed that traditional herders have less access to fattening facilities and also, there are many native breeds that are less responsive to intensive feeding as compared to the improved ones.

6.9. SOCIAL, ECONOMIC AND ENVIRONMENTAL FACTORS

It has been observed that often the issues of grasslands and their users are more socio-economic as compared to that of technical. Better management and enhanced livelihoods can only be obtained if the legal, social and economic problems that are related to pastoralism, are dealt with.

There is the recommendation of "Training herders" in many of the projects, but it is pointless to try to transfer technical ideas, probably developed elsewhere, when the herders have no safety of tenure, the techniques and approaches have not been convincingly tested locally. In addition to it, poverty and population pressure mean that herders will not take any kind of extra risk.

6.9.1. Tenure

Secure tenure of land or grazing rights is very important if stock-raisers ,as well as pastoralists, are to have safe livelihoods and can pay money in and manage grasslands in a sustainable fashion. The place where the grassland production systems are purely commercial (as in the studies on the commercial sector in South Africa, Patagonia, the Campos and Central North America) the land is held in either freehold or it is held in long-term leasehold.

Therefore, the commercial stock raisers can capitalize on infrastructure, mainly water and fencing – a primary use of fencing may be to delimit properties. Since it has been observed that commercial enterprises hold valid land titles their land can be further used as collateral for finances or loans.

It has been observed that in the wide-ranging, pastoral subsistence sector, grazing rights are much less clear. In the distant past, these kinds of lands would have been handled under traditional authorities and disputes over encroachment by some other herding groups or cultivators that are probably settled in battle.

The land rights that are for settled farmers are identified in most of the countries since they are resident on their farms and obviously make use of them; the rights of pastoral groups, though, who are generally mobile, are usually less well defined since they only make use of a piece of grassland at a specific season.

If some people clear such grassland for crops, however unsustainable, it may be seen as "development" and it has been observed that pastoralists are at a drawback in claiming their rights. In addition to it, traditional pastoral tenure is not generally strong enough to stop confiscation by the state, probably without compensation, for infrastructure, mineral prospection, building or nature reserves.

While cropland can suitably be assigned to individual smallholders, the large areas of low-yielding grassland that are included in mobile herding and the desirability of managing such kind of pasture at the landscape scale make the distribution of grassland to individual families problematic (however such kind of allocation has been done in China).

The allocation to groups seems to be preferable, but at what scale and how to decide to whom grazing should be assigned is difficult.

6.9.2. Markets and Trade

Of course, commercial systems are market- oriented, and in the present times, most of the traditional systems sell their surplus production. The influence of freer world trade on produce from widespread grasslands has still to be seen, but meat that is produced by traditional herders who are far from the consumers may be at a disadvantage – particularly as it has been observed that urban consumers increasingly need meat from cattle that have been finished in feedlots or off good pasture.

6.9.3. Herder Organization and Community Participation

Old herder groupings and hierarchies have been disturbed by the regime and political change. In addition to it, de-collectivization has left large areas with a disorganized pastoral sector.

Herds are often too small in size for it to be profitable for all the families to spare labour to take them to pasture, and herding communities are fragmented.

It has to be noted that some planning is required at the landscape scale if semi –arid grasslands are to be managed sustainably. In the present times, it is widely accepted that rural development, which includes grassland development, should be led by the ultimate users.

Community participation is very important, but if it is to be effective, in spite of token talking, a high priority should be the development of some ways of having herders form themselves into larger groups for deciding local herding policy, discussing the policies with the regional authorities and sharing all the herding tasks.

6.9.4. Diversification

It has been observed that a number of commercial enterprises in most of the countries studied are looking at alternative as well as potentially more profitable uses for their grasslands. This is because of poor returns from animal husbandry.

Raising game and wildlife is already practised in eastern and South Africa and is probably expanding – this may be for specialist meats, tourism, hunting of a combination.

Some mention organic meat

Tourism and eco-tourism are another use of grasslands; in commercial region its benefits will go to the landowner; the extent to which it will become important is not quite clear except where significant scenery or wildlife is involved from the time, several numbers of grassland regions are remote and have little infrastructure.

Tourism is motivated with the help of several numbers of governments since it brings them revenue. Never the less, in the regions of traditional herding, tourism must be seen as profitable by the graziers who take participation; the owners of large private establishments may negotiate fees, but in traditional systems, tourists may be regarded as a nuisance if they make no contribution to local livelihoods.

6.9.5. Grassland in The Environment

Even though the grasslands are of major environmental significance, not least as catchment regions and sites for in situ conservation of biodiversity, their preservation along with the appropriate management is given comparatively little attention with the help of environmentalists and the governments. That as an outcome, every so often see the traditional livestock sector more as an issue as compared to the important part of maintaining grasslands as well as their biodiversity.

There are several numbers of reserves and national parks and they are increasing in numbers as well; they every so often decrease the conventional grazing lands, with little or no attention to their traditional users. These kinds of reserves are for wildlife, biodiversity and sometimes the consequent tourism, but the grassland biome of these types of reserves necessitates accurately managed grazing with respect to its survival. Grassland reserves as such are rarely mentioned: China (Hu and Zhang 2003) has eleven, covering two million hectares.

6.10. REHABILITATION OF DEGRADED FORESTS IN ASIA

1. Grasslands in Asia can be categorized according to origin.
2. Grasslands caused with the help of climatic or edaphic reasons; and
3. Grasslands caused by the conversion of forests with the help of cueing or burning.

The total region of grasslands in Asia is not well documented, but in the year 1987, according to the reports of FAO, that estimated it to be 537 million hectares. How much of this is human -induced is not accurately known either, but some indicative figures are available for some countries.

In Table 6.1. that is mentioned below, provides data for selected nations where the grasslands spread over the large regions. These several numbers of nations are consisting of India, China, Indonesia, the Philippines, Thailand, and Sri Lanka

Table 6.1. The extent of grasslands in selected countries in Asia (thousands of hectares)

Total Land		**Area Under**
Country	Area	Grassland
China	932,641	279,792b
India	297,319	30,000'
Indonesia	181,157	12,000
Pakistan	77,088	5.000
Philippines	29,817	7,000d
Sri Lanka	6,474	64'
Thailand	51,089	2,000'

Source: Table from nzdl.org

6.11. DESCRIPTION OF HUMAN-INDUCED GRASS-LANDS

Grasslands are dominated with the help of particular kind of groups of species that relies on the amount of rainfall and temperature grades as well as on the degree of biotic pressure and disturbance.

In several regions of the Philippines, there are four vegetative associations that have been recognized (Singh, Hanxi and Sajise 1985).

The four vegetation associations are mentioned below:

1. *Imperata cylindrica* community with *Heteropogon contortus*, *Chrysopogon aciculatus*, *Manisuris clarkei*, *Paspalum dilatatum*, and *Capillipedium parviflorum* as the main associates;

2. *Themeda tirandra* community with *C aciculatus*, *P. dilatatum*, *I. cylindrica*, *C. parviflorum* and *Coelorachis glandulosa* as the main associates;

3. *Chysopogon acicu/atus* community with Ttriandra, *I. cylindrica*, *P. dilatatum*, *Alysicapus vaginalis* and *Fimbrisylis monostachya* as important species; and

4. *Capillipedium parviflorum* community with *I. cylindrica*, *P. dilatatum*, *Heteropogon contortus*, *F. monostachya* and *C. acicu/ atus* as main associates. This last community is associated with overgrazing.

With respect to India, there are five major types of grass cover have been found, that are mentioned below:

1. Schima-Dicanthium type;
2. Dicanthium-Cenchrus-Lasiurus type;
3. Phragmites-Saccharum-Imperata type;
4. Themeda-Arundinella type; and
5. temperate alpine type. The first is tropical and is less extensive.

The second, third and fourth are subtropical. The Themeda-Saccharum-Imperata type, which is akin to the Imperata grasslands, is found in the moist and wet habitats (Singh, Lauenroth and Milchunas 1983).

In China, seminatural grasslands are predominantly *Imperata cylindrica*, *Paspalum scorbiculatum*, *Cymbopogon caesius* and species of Aristida, Ischaemum, Heteropogon, Eulalia, Digitaria, Eragrostis, and Arundinella (Singh, Lauenroth and Milchunas 1983).

6.12. CONCLUSION

With the grasslands getting highly impacted with the severe changes in the climate, it is very important in the present times to rehabilitate and improve the grasslands. Not only do the grasslands are the most important of the biodiversity, the grasslands even cover a large portion of the world.

Thus, in order to improve the grasslands, it is very important to revive and maintain nutritional quality. Along with maintaining the nutritional value of the plants, it is also very important to maintain the fauna that dwells in the grasslands.

Some of the researchers state that the nutritional value of the grasslands does not focus only on increasing the nitrogen content in the plants. This is so because excessive nitrogen in the plants decrease the silage in the grasslands. This, in turn, brings down the nutritional health of the grasslands.

The methods discuss in this chapter could prove to be very helpful in improving and rehabilitating the grasslands along with improving the social, economic and biological aspects of the world.

REFERENCES

1. Chaplot, V., Dlamini, P. and Chivenge, P. (2016). Potential of grassland rehabilitation through high density-short duration grazing to sequester atmospheric carbon. Geoderma, [online] 271, pp.10-17. Available at: https://www.sciencedirect.com/science/article/pii/S001670611630060X [Accessed 28 Feb. 2020].
2. Duthie, D. (1955). Grassland Improvement. The East African Agricultural Journal, [online] 20(3), pp.137-139. Available at: https://www.tandfonline.com/doi/abs/10.1080/03670074.1955.11664954?journalCode=teaf19& [Accessed 28 Feb. 2020].
3. Food and Agriculture Organization (n.d.). Grassland of the world. [online] Fao.org. Available at: http://www.fao.org/3/y8344e0j.htm [Accessed 28 Feb. 2020].
4. Food and Agriculture Organization (n.d.). Plant Production and Protection Division: How to management grassland and pasture areas. [online] Fao.org. Available at: http://www.fao.org/agriculture/crops/thematic-sitemap/theme/compendium/tools-guidelines/how-to-management-grassland-and-pasture-areas/en/ [Accessed 28 Feb. 2020].
5. Kemp, D., Guodong, H., Xiangyang, H., Michalk, D., Fujiang, H., Jianping, W. and Yingjun, Z. (2013). Innovative grassland management systems for environmental and livelihood benefits. Proceedings of the National Academy of Sciences, [online] 110(21), pp.8369-8374. Available at: https://www.pnas.org/content/110/21/8369 [Accessed 28 Feb. 2020].
6. Leafe, E. (1988). Introduction—the history of improved grasslands. The Grass Crop, [online] pp.1-23. Available at: https://link.springer.com/chapter/10.1007/978-94-009-1187-1_1 [Accessed 28 Feb. 2020].
7. Mingyi, F., Zhong, Y., Yulin, D. and Jinfeng, H. (2003). Grassland Rehabilitation and Social Development in the Low Mountain Area of the Jinsha River Valley, China. Mountain Research and Development, [online] 23(2), pp.124-127. Available at: https://bioone.org/journals/mountain-research-and-development/volume-23/issue-2/0276-4741(2003)023%5B0124%3AGRASDI%5D2.0.CO%3B2/Grassland-Rehabilitation-and-Social-Development-in-the-Low-Mountain-Area/10.1659/0276-4741(2003)023 [0124: GRASDI]2.0.CO;2. full [Accessed 28 Feb. 2020].

8. Muller, S., Dutoit, T., Alard, D. and Grévilliot, F. (1998). Restoration and Rehabilitation of Species-Rich Grassland Ecosystems in France: A Review. Restoration Ecology, [online] 6(1), pp.94-101. Available at: https://onlinelibrary.wiley.com/doi/abs/10.1046/j.1526-100x.1998.06112.x [Accessed 28 Feb. 2020].
9. Nzdl.org. (n.d.). Rehabilitation of Imperata grassland. [online] Available at: http://www.nzdl.org/gsdlmod?e=d-00000-00---off-0hdl--00-0----0-10-0---0---0direct-10---4-------0-1l--11-en-50---20-about---00-0-1-00-0--4----0-0-11-10-0utfZz-8-00&cl=CL2.18.2&d=HASH0f4b24345cce2d6306addf.8.5&gt=1 [Accessed 28 Feb. 2020].
10. YARA (2020). How to improve grassland quality | Yara UK. [online] Yara United Kingdom. Available at: https://www.yara.co.uk/crop-nutrition/grassland/improving-grassland-quality/ [Accessed 28 Feb. 2020].
11. Prof. Dr. Martin Elsaesser (2014). Grassland renovation and grassland improvement. [ebook] Germany: LAZBW Aulendorf. Available at: https://web2.mendelu.cz/af_291_projekty/files/19/19-elsaesser_grassland_renovation.pdf [Accessed 28 Feb. 2020].
12. The James Hutton Institute (2020). Improved Grassland | Land Capability for Agriculture | Exploring Scotland | The James Hutton Institute. [online] Hutton.ac.uk. Available at: https://www.hutton.ac.uk/learning/exploringscotland/land-capability-agriculture-scotland/improved-grassland [Accessed 28 Feb. 2020].
13. Moore 1964
14. Buringh and Dudal 1987
15. Elsaesser, 2005
16. Taube and Conjin 2007
17. Soegaard 2007
18. Hu and Zhang 2003
19. Singh, Hanxi and Sajise 1985
20. Singh, Lauenroth and Milchunas 1983

7

Effect of Grassland Management on Plant Production

CONTENTS

This chapter explains the importance of plant species in the environment and importance of maintaining them. This chapter explains the types of plants that grow in the grassland. As these plants need different conditions for their better growth, so these conditions are also being explained.

With this, the reasons that are responsible for destroying this grassland vegetation are also discussed. The importance of vegetation in the society is very vast, so maintain them in a proper manner in a given guidelines is very important. This chapter also discusses the different factors that are responsible for bringing the loss of vegetation.

7.1. INTRODUCTION

Grasslands are considered important components in the field of agrarian landscapes. As grasslands provide a wide range of ecological, economical, and social goods and services that are necessary for the support life. Grasslands are found to provide a habitat for a numerous plants and animals.

They also help in filtering sediments and pollutants before they reach to the freshwater sources. Grasslands also helps in increasing water infiltration, prevents soil erosion, they also help in removing and storing the greenhouses gases.

Plant communities are considered as the important and main element of the grassland ecosystem.

The ecological conditions of the habitats are reflecting conditions of Plant communities and species composition. These two factors are considered as an important bio indicators. Ecological factors are the important factor that is required by the grassland plants.

The ecological factor is considered as most suitable for them and on the other hand they are optimal to growth and development. However, these economical activities that are carried out in the grasslands gives a negative impact on the relation of the ecological conditions and to the plants. There are some long-term changes in the condition of the habitat which are determined by the various ecological factors.

These long-term changes make an impact on the condition of the plant and competitiveness. Therefore, with this the species composition of the communities also gets changed.

Extensive Grasslands are said to have multiple uses. It is true that they are a very important source of the livestock feed and livelihoods for the raisers and herders. Minahi et al. in the year 1993 states that grasslands are as important as forests. As grasslands help in recycling the greenhouse gases and the soil organic matter.

7.2. IMPORTANCE OF CLASSIFICATION OF PLANTS

Grasslands are very useful due to the present of different types and wide variety of plants that are found in them. The main feature of the grassland agriculture totally depends on the types of grassland plants.

There are different plants types that are increasing the benefits to the livestock and wildlife. To have a proper grassland management, it is important to have the knowledge of various types of plants that are present in the grassland.

7.3. GRASSLAND PLANT CLASSIFICATION

Grassland includes many different types of species of plants. Grassland plants are classified according to the two different criteria. These two criteria are: life cycle and plant type.

On the basis of the life cycle, the plant species are divided based on their yearly growth and characteristic of the seeding. Each plant grows in stages and these stages together make a life cycle.

Plant cycle is really important and their life cycle is related to the productivity of the plant. Therefore, the plant can be divided on the basis of their life cycle. According to the life cycle, plants can be categorized into three categories and that are: annuals, biennials, and perennials.

Annual plants: Annual plants are those plants that complete their life cycle with one year or growing year. these types of plant die after the production of seeds. They do not grow next year until they are being planted again. Examples of such plants are corns, crabgrass, sunflower, etc.

Biennial Plants: Biennial plants are those plants that require two years to complete their lifecycle. These plants are generally producing vegetative growths that will include leaves, stems and roots.

This process takes place in the first year. After that year, these plants produce flowers, fruits and seeds in the next year or second year. After completing their two years, in the end they die. An example of such a biennial plant is Red clover.

Figure 7.1. Annual and Biennial plant.

Source: Image by Wikimedia Commons

Perennial Plants: Perennial plants are considered to grow year after year. Perennial plants produce flowers, fruits and seeds every year. After the production, these pants go into the resting phase known as dormancy.

This process of dormancy is best noticed in the areas where the weather is cold. In these areas these plants stop growing completely for the winter season. During the period of dormancy, these perennial plants slow down their natural process in order to protect themselves from the cold temperature.

When the spring season comes, these plants start growing again. Examples of perennial plants are forages and pasture crops. Trees and shrubs are also considered as perennial plants.

Plants are basically categorized according to the physical characteristics of the plants. Where each of the species within the plant type is being characterized that is similar to all the other species of the same kind. There four different types of grassland that are found and that are grasses, legumes, forbs and woody plants.

7.3.1. Grasses

Grasses are considered as one the four types of grasslands that are commonly found. Grasses are considered to serve many purposes. These purposes basically include food for livestock and wildlife that includes orchard grass, fescue and alfalfa, food for humans that may include cereals and grains, and these grasses can play a role in erosion prevention.

The grasses that are present on the Great Plains are basically considered as herbaceous or plants without woody stems. The grasses basically resist compaction because the stems and culms of the grasses are usually hollowed. Leaves or blades get connected directly to the culm at a sheath. These leaves are basically surround the stems.

All the leaf blades basically have a distinctive parallel venation. In this venation the veins run side by side along the length of the blade of grass. This characterisation of visual makes possible to separate the grasses from all the plant species.

Figure 7.2. An image representing Grasses, type of grassland.

Source: Image by Flickr

The grasses are divided into two major groups, these groups are cool -season grasses and warm -season grasses. Cool -season grasses tend to grow best during the spring and fall season.

These grasses plants begin their growth when the soil temperature reaches the temperature of 40° F. But these plants' optimum growth starts when the air temperature falls between 59° to 77° F .

These plants can remain green all the winter, but they tend to get brown and dormant during the summer season. These plants can be considered under annual as well in perennial. Examples of cool- season grasses are Kentucky bluegrass, orchard grass, smooth brome grass.

Whereas, warm- season grasses are the opposite of this. These plants grow best in the summer season or in warm temperature. These grasses are considered to be much more tolerant toward to heat and drought than cool -season grasses.

These grasses have a growing season that begins when the soil temperature reaches 60° F. these plants are considered to grow best during the summer season, at that time the temperature ranges from 77° to 104° F.

These summer season plants become dormant in the winter season and they do not start growing until early summers come. They can be count as perennial as well as annual plants. Examples of these plants are Indian grasses, big blue stem and switch grass.

7.3.2. Legumes

Legumes and grasses together make the dominant plant type that is commonly found in the grasslands. Both of these types of plants are used as a forage crop. These plants are very beneficial to agricultural production.

Figure 7.3. Types of Legumes.

Source: Image by Wikimedia Commons

Example of legumes plant basically includes, soybeans, alfalfa, coves, birdsfoot trefoil, etc. Legume plants are classified using different characteristics. One of the characteristics of the legumes plants is a fruit or pod that they produce. These pods are basically containing only one chamber with the seeds that lined in a single row. Different plants have different pod size and seed numbers.

All the legume has leaves. These leaves are arranged alternatively on the stem. These leaves are connected to the stem by a stalk known as petiole. It not like in grasses, legumes contain venation that consists of a network of veins, instead of veins that run parallel to each other.

Legume plants can consider as annual, perennial or biennials. Most of the legumes have the ability to take the nitrogen from the air between the soil particles. After taking nitrogen, these legumes change it into a form of nitrogen that can be used by the plants.

This process is known as nitrogen fixation. This process is basically carried out by the symbiotic bacteria that are found in the nodules on the roots. This nitrogen helps them to decrease fertilizer need, reduces the cost, increases the yield and also enriches the soil.

7.3.3. Forbs

Forbs plants are those that are not considered as grasses neither legumes, those plants are called as forbs. Most of the forbs are broad -leafed. The broad-leafed structure makes possible to distinguish them from grasses and grass like plants or legumes. Forbs plants don not usually cultivated for agricultural production. They commonly appear in both pastures and native plant habitats.

Many of the forbs are valued as wildlife food and they are basically used for covering the soil from erosion. Whereas, other forbs are considered as a noxious weed. Forbs can be considered as perennials, annuals or biennials. For example, these forbs plants are sunflower, ragweed and thistle.

Figure 7.4. Image of a forbs plant known as ragweed.

Source: Image by Needpix.com

7.3.4. Woody Plants

Woody plants are those plants that are considered as the easiest plant that can be identified in the grassland. These plants contain tough, woody stems commonly known as non- herbaceous. Woody plants can be either shrubs, vines or trees.

Figure 7.5. Image of a Red Cedar woody plant.

Source: Image by Wikimedia Commons

Woody trees are basically found in the grassland. Due to this reason, these plants are always found to be immature due to the nature and use of the grassland. These woody plants are kept small by the animals as they graze their terminal branches.

These woody plants are small because of the fires that stunt the growth, by mechanical cutting or by the chemical treatment that is carried out to maintain the grassland. Woody plants are considered as a perennial plant. Examples of such woody plants are wild roses, elm and red cedar.

7.4. ENVIRONMENTAL BENEFITS OF GRASSLANDS

Grassland is considered to be an important component of the agrarian landscape for distinguishing itself. This process of distinguishing is considered to have multifunctional roles which help in the formation of landscape.

As, grassland performs some essential services that are necessary to support life and also contributes to human well-being. It also provides some beneficial goods and services that are important for extending the local, regional and global communities. The main function of the grassland is to provide the feed for the livestock, but it is considered that grassland also serves other functions that are important.

The habitat of grassland is considered to be very important in storing the biological varieties and the components of the agrarian landscape. These habitats of grasslands are having probably the largest biological variety. The grasslands that are present in Europe are said to contain the diversity of plants, insects like butterflies, birds or fungi that are basically exceptional (Hönigová et al., 2012).

A grassland is considered to contain up to several dozens of the plant species that may be present with the one square meter of area. There are some varieties of grassland plants that play an important role in the economic, ecological, aesthetic and unmistakably and formation of landscape meaning.

Grassland plays an important role in water circulation that is considered as a very important role of grasslands that helps in maintain nature.

For the enlargement of the air humidity, forests together with the grassland play an important and significant role (Jankowska-Huflejt, 2006). Whereas, grassland is an important traditional source that is used as a medical plant and used for other medical resources.

The pharmaceutical uses of medicinal and aromatic plants are being connected with the active substance. These active substances can be oil or tannins (Hönigová et al., 2012). It should be noted that the natural mechanism of flood regulation is considered as one of the important ecological functions which can be performed by flooded grasslands. Therefore, the ecosystem influences the timing and magnitude of the water runoff, flood and aquafer recharge and this is defined as the grassland services which is said to be important for water regulation. These grassland services of water regulation can be explained particularly in terms of water storage potential of the ecosystem (Hönigová et al., 2012).

With all the above functions, a grassland also performs an important function that works in decreasing the amount of gas that can cause the greenhouse effect in the atmosphere and in the solution of ecological problems that is in the global context. Grasslands are being considered to have the potential that can absorb the carbon that is present in the atmosphere.

Thus, grasslands contribute to the reduction of the amount of the main gas that can be responsible for causing the greenhouse effect, specifically CO_2. Therefore, grassland acts as a significant carbon sink with the implementation of improved management (Conant et al., 2001).

7.5. MAIN DRIVERS INFLUENCING THE GRASSLAND DEGRADATION

There are several mechanisms that are responsible for grassland degradation. These mechanisms are highlighting climate change and anthropogenic disturbances. There are basically two main factors that are underlying the fact that lead to degradation.

Altogether, these factors of climate change are considered to be most related to the grassland degradation that are basically warming and the increase in the frequency is the prolonged drought. There are some researchers like Gang et al., Zhou et al., Yang et al., and Liu et al working on presenting the quantitative assessments of the relative roles of climate change.

There are some other roles like anthropogenic perturbations, and they can also be used for revealing that any of them can works as a dominant driver in degradation that depends on the study of the context. Therefore, most of these studies did not consider the interaction that takes place between the drivers.

In this case, Sala et al. in the year 2017 show that there is clear evidence that helps in supporting non-linear responses of the ecosystem. They also help in synergistic interactions that are among many of the global drivers of biodiversity change.

For example, invasion of the alien species that are basically prompted by both human disturbances and climate changes.

Klein et al., in the year 2007, on the other hand reveals that there is a decrease on the ANPP and forage quality for 1.0 to 2.0 °C warmer. Klein also showed that an interaction with the grazing could mitigate that can decline.

With addition to this, Carley et al., in the year 2014 analyzed that there are different grassland types that are found by manipulating the temperature, water and grazing conditions. These conditions explain that biomass production depends on the interaction that takes place between the three variables that comes under the site-specific conditions.

Overgrazing basically occurs when the stock rates exceed the carrying capacity of the grasslands. Overgrazing is considered as the most widespread anthropogenic driver responsible for degradation.

In fact, it is considered that stocking rates are comparatively less variable than carrying capacity. Which is ultimately related to the primary production and to the climate conditions.

This is the way which helps overgrazing to clearly interact with climate variability.

Lui et al. in the year 2013 report on this regard. Lui explained that the Mongolian steppe degradation can be attributed in the context of both climate change which involves decreasing precipitation and increasing the temperature and with this it also attributes to the increase in the stocking rate.

There are some more factors that are working as a source of degradation that are affecting the productivity and species composition. These factors are extreme under grazing or grazing abandonment.

There are also some of the long term livestock enclosures that are responsible for increased standing biomass and vegetation cover.

Although, they may be useful as a management tool that works to restore the vegetation in the degraded grassland of arid regions.

They may sometimes, decrease the grassland biodiversity and ANPP. In regards to grassland degradation, there is a long-standing controversy that is about the use of fire.

Snyman in the year 2015 suggested that fire can potentially work in degrading the semiarid grassland.

This degradation is based on the fact that the recovery of the above and below ground biomass production and water use efficiency takes at least two years after burning.

On the other hand, when the fire is used, then fire appears to be relevant in the fire- prone communities, for example in the case of mesic grasslands. In this, fire is an evolutionary force that has been used for hundreds of years in order to maintain the grassland structure and function.

Finally, there is improper fertilization and introduction of exotic species and there are also other important factors like invasive of alien species that may influence the grassland degradation.

7.6. IMPACT OF HUMAN ACTIVITIES ON THE CHANGE OF GRASSLAND PLANT COMMUNITIES

There are some long term changes in the conditions of the habitats that include chemical and physical properties of the soil, conditions for irrigation, etc. the long term changes are determined by the various ecological factors that may influence the conditions of the plants and their competitiveness. Therefore, later the composition of the species of the communities gets changed (Svirskis, 2004).

When there is a change in the quality of the plant communities then it effects on all the forms of life. The conditions of the plant communities and species composition basically reflect the ecological conditions of the habitats. They also serve as an important bioindicators (Karpavičienė, Marcinkonis, 2009).

An anthropogenic factor is responsible for the great impact on the formation of the grassland land plant communities. As, there is no grassland present that is completely protected from human activities. The communities of the grasslands are not considered as constant and the changes are due to such human activities.

These activities basically include, burning, drainage, addition of the seed of herbs, non-usage of grasslands, drainage, contrariwise, creation of sown grassland, haymaking and grazing. There are certain factors that are included in the human activities that are responsible for effecting the grassland plant communities are described below:

7.6.1. Fire

These types of practices are commonly known by officials as dangerous because they can even lead to fire of the building and forests (Russel et al., 2009). Burning in many countries is applied as a means of environmental protection. There are many nature management schemes which are being evaluated ambiguously as the application of burning.

There are some scientists and specialists working for protecting nature, do not approve the burning. They also motivate that this method destroys the plants and animals, pollutes the environment. They think that it is not a proper suspension for the succession processes.

Fire may create monocultures of the fire-resistant species in the degraded or low productivity grasslands. There are some cases where fire may work as a significant promoter of the exotic species. There are some other scientists who believe that the controlled burning is necessary for seeking to sustain the development of the grasslands. This development is said to be formed historically by Harrison et al., in the year 2003.

The other ecological factor that is affecting the development of the variety of the grasslands is flame. Flame is considered as an important factor of the environment. Most often there are some anthropogenic origins that are responsible for destroying and forming the ecosystem at the same time. Fire is said to occur very often in the grasslands of the temperate climate zone.

People think that the burning of the grasslands helps in improving the soil, regenerates the soil and fertilizes it. The action of the fire modifies the soil environment and therefore changes in the biological activity of the soil that may be expected following burning (Russel et al., 2009).

7.6.2. Fertilization

The most important sign that is used for the communities is the species composition of the grassland plants. There are many scientist researches that indicate that amount of nitrogen and phosphorous increases with the use of mineral fertilizers. Thus, there is a decrease in the variety of grassland plants (Schellberg, Pontes, 2011).

There are only a few species of plants that grow better in theee soil which gets saturated with the nutritional substances. Example of such plants which are considered as the indicators of the soil which is basically rich in the nitrogen.

These plants are nettles (*Urtica dioica*) meadow foxtail (*Alopecurus pratensis*), perennial ryegrass (*Lolium perenne*), orchardgrass (*Dactylis glomerata*), cow parsley (*Anthriscus sylvestris*), (Karpavičienė, Marcinkonis, 2009).

There is some largest variety of vegetation that is basically observed in the grasslands which may contain a small amount of nutritional substance (Plantureux et al., 2005). When there was a decrease in the species varieties of the plants is being explained by the different abilities in order to compete for the light.

As light is considered as one of the most important ecological factors that is influencing the growth of the plants. While in the competitive fight for the light, plants that are growing taller and are most rapidly growing are considered to have an advantage (Pärtel et al., 2005).

In the case of grasslands that get saturated with the nutritional substance, those grasslands are considered to have the relative differences of growth which increases even more. There are several species of plants that can compete for the light, which can be overshadowing other plant species.

These species of plants start gradually to dominate other plants that are able to compete for light and those plant species which eliminate the lesser competitive species. This basically decreases the variety of grassland vegetation (Pykälä, 2007).

There have been much researches that is done on fertilizers which have shown that the species composition of the grassland plant communities is highly dependant on their fertilization. It has been established that even the amount of used fertilizer is very low, the significant role of the number of plant species is being observed. Even when it is compared with the amount that is used in intensive agriculture.

Also, it is very often that the usage of the nitrogen which is supplementing the grasslands during fertilization is inefficient. The fertilizer that is used can be mineral fertilizer or it can be manure.

It is the large amount that is washed out or amount that enters the atmosphere. In order to seek decrease in the negative impact on the environment and economic losses, it is very important to establish the amount of nitrogen that is present in the fertilizer.

This fertilizer basically assimilates the grassland plants most effectively. While fertilizing in various intensiveness it is important to note what amount of the nutritional substance is washed out from the grassland soil or what amount does gets evaporated to the atmosphere. It is important to note how different methods are used to influence the loss of nitrogen.

For example, in the case of small outflow, it is being established during the scientific research that the smaller amount of nitrogen enters the underground water basically when the grasslands are trimmed and being grazed when compared to the only grazed ones (Jankowska-Huflejt, 2006).

7.6.3. Haymaking

For the protection of the biological variety of the grassland ecosystem, the peculiarities of the haymaking are considered as important. These peculiarities basically include the number of harvesting season, time of haymaking, the method of the management of the cut grass, and the height of the cut plants.

The direct influence of the haymaking is displayed by the rapid changes that take place in the rhythm of the vegetation. This display also takes place by the accumulatio of the nutritional substances, maturation of the seed which gets unbalanced; therefore, plants are able to adapt the rhythm of the haymaking that starts to dominate.

Figure 7.6. An image representing haymaking.

Source: Image by Pixabay

After the haymaking, phytoclimatic changes are rapid when the energy of the sun reaches the soil and less amount of plant residues gets accumulated in the soil. Thus, the soil gets heat up better and dries and with this there is improvement in the movement of gas that takes place between the soil and the atmosphere.

The haymaking of the grasslands may consider as an important factor to the principal component of the communities (Rimkus, 2003). It has been scientifically proved, that the mowed species varieties of grassland vegetation are larger when compared to the grasslands which are mowed two or more times within the season.

Therefore, the grass with frequent usage is considered of better quality. These grasslands contain many indicators of intensive farming. They are also said to contain nutritional substances, dry substances and digestibility of the organics, palatability and other properties (Vasiliauskienė et al., 2007).

However, in the year 1998 – 2006 after the evaluation of the material of the investigations that is performed in the territory of the Nemunas Regional Park.

7.6.4. Grazing Heavy

Grazing also influences the communities of the grassland in the negative form. During grazing as well as during mowing, it is considered that the natural development of the grassland is damaged. However, the grazing effect is different from the mowing effect, which means that the grazing affects the botanical diversity differently. This effect is done by selective defoliation that is due to dietary choices, trading and nutrient cycling.

In order to maintain the vegetation typical to the ecosystems of the grasslands is done by grazing at low stocking rates is especially important. It also prevents it from covering with bushes and trees.

There are basically two factors that are main to explain the plant species richness and activities that are related to the grazing activities. These activities are the stocking rates that is the density of the animal per hectare and it is the duration of the re-growth that takes place between the grazing period.

When there is a use of grassland for the grazing process, the stock should not exceed 1.5 animals per hectare. The re-growth duration should be given at least 35 days, this is done so as the species variety of the vegetation should not decrease.

It is considered that the grazing which is done intensively, produces short dense swards.

These swards generates low amount of seeds. Inversely, if there is an intensive grazing then that will provide the conditions for the grasslands grass to grow high and lush (Plantureux et al., 2005). In the year between 1993 to 2002, the Institute of Agriculture present in Lithuanian is said to perform a long research in Dotnuva.

Their main aim was to establish the impact of extensive grazing that happens on the vegetation of the grassland. Before the year that is mentioned above for the research, another investigation was also performed in the year 1961 to 1992.

7.6.5. Melioration Multiplex

Melioration of the grasslands helps in determining that there is a spread of new types of plants in these grasslands. When the grasslands get drained then the types of plants that are typically to the wet grasslands and even there is a larger decline of the biological variety which is observed in these ecosystems and they are used more extensively.

The research conducted in France during the scientific experimental research showed that these species compositions of the grasslands get drained for the periods ranging from 1 to more than 30 years when compared.

Figure 7.7. Image representing the formation of Melioration channels.

Source: Image by Wikimedia Commons

It is being established that the dominance of the plant species that is being modified during the growing season. This is flowed by the drainage and that specific species gets disappear or appeared, which significantly changes after five years (Isselstein et al., 2005; Plantureux et al., 2005).

7.6.6. Abandonment

There are some plant communities of the grasslands that are not stable. With this the vegetation responds to the environmental conditions. There is some transformation that takes place in the grasslands into arable lands or reseeding that is mostly in the cultural grasses.

This transformation mostly influences communities. However, the abandonment of the grasslands that also has a negative effect. This negative effect is due to the succession process that changes the grassland plant communities.

In such cases, there is a considerable threat of the spread of the grasslands which are bushing. This process takes place with the alders (Alnus incana) and the young forest Betula pendula and Salix bushes, arises. The grasslands after overgrazing or mowing get terminated in the abandoned grassland. Plant communities of the grasslands during the succession grow over with the woody plants.

When the natural succession of the plant towards the forest habitat formation takes place, in that condition the natural habitat of the environment of the plant species vanishes. As agriculture is considered to have a lot of influences that take place over the course of succession. The succession of the communities is said to move and grazed grasslands, on the other hand, is constantly influenced.

Therefore, there are some locations where forests get formed naturally. It is also believed that there is an ecosystem of grasslands that are retained. In Lithuania, all the grassland plant communities are expected to flood the grasslands of the river valley. This happens due to the conditions of the local climate that are related to human activity.

Therefore, when there is an active economic activity that is being terminated in these territories then the factors are expected to get suspend succession that vanishes and there is a formation of the forest habitation (Isselstein et al., 2005).

However, this process of succession can also be favorable for the restoration of botanical diversity. When there are abandoned and ex-arable lands then that gets gradually covered by the vegetation. This vegetation is basically the grasslands or the species that are composed of sown grasslands. They are even believed that they get close to the communities of the natural grasslands.

7.7. CONCLUSION

This chapter explains how grassland management is important for maintaining plant production. There are many plants that are present in the types of grassland plants. These plants are of different type that grows according to the area and the climate conditions.

Figure 7.8. Present of different types of grassland plant.

Source: Image by Geograph

Like there are some plants that grow even when there is excessive cold and some plants grow even when the temperature is too high. This simple shows that all the plants of grassland need different conditions for their growth.

There are different methods like water management, Salt licks, fire, bush control, fencing. These methods are being used to control the grassland production. These methods even help in developing plant growth in the land. Even some methods are being used to protect the plant species form grazing.

There are some effects that are being caused by human beings that are destroying the vegetation of the grassland. These factors are being explained in this chapter. These factors are fire, fertilization, haymaking, heavy grazing, etc.

Human beings that are not following the standards that are being used to maintain the grassland vegetation are the main cause of destruction. As human beings are using fire for developing the new varieties of plant species and to maintain the soil texture. If they are not using guidelines for the fire, then that fire can cause great destruction. That can be, vegetation loss, animal loss and forest loss, etc.

So, it is important to maintain the grassland plant production, but one should use the correct method that can be used to maintain them in a proper manner.

REFERENCES

1. F, R., Miller, D. and Nelson, C. (1995). *Introduction to Grassland Management*. 5th ed. [ebook] Columbia. Available at: https://dese.mo.gov/sites/default/files/aged-Grassland-Student-Ref..pdf [Accessed 25 Feb. 2020].
2. Fao.org. (n.d.). *Plant Production and Protection Division: How to management grassland and pasture areas*. [online] Available at: http://www.fao.org/agriculture/crops/thematic-sitemap/theme/compendium/tools-guidelines/how-to-management-grassland-and-pasture-areas/en/ [Accessed 25 Feb. 2020].
3. Heyburn, J., McKenzie, P., Crawley, M. and Fornara, D. (2017). Effects of grassland management on plant C:N:P stoichiometry: implications for soil element cycling and storage. *Ecosphere*, 8(10), p.e01963.
4. Ignatavičius, G., Sinkevičius, S. and Ložytė, A. (2013). Effects of grassland management on plant communities. *Ekologija*, 59(2).
5. Tiscornia, G., Jaurena, M. and Baethgen, W. (2019). *Drivers, Process, and Consequences of Native Grassland Degradation: Insights from a Literature Review and a Survey in Río de la Plata Grasslands*. [ebook] columbia. Available at: https://www.mdpi.com/2073-4395/9/5/239/pdf [Accessed 25 Feb. 2020].
6. Carlyle et al. [2014]
7. Conant et al., 2001
8. Harrison et al., 2003
9. Hönigová et al., 2012
10. Isselstein et al., 2005
11. Jankowska-Huflejt, 2006
12. Karpavičienė, Marcinkonis, 2009
13. Klein et al. 2007
14. Minahi et al. in the year 1993
15. Pärtel et al., 2005
16. Plantureux et al., 2005
17. Pykälä, 2007
18. Rimkus, 2003
19. Russel et al., 2009
20. Sala et al. [2017

21. Schellberg, Pontes, 2011
22. Snyman [2015]
23. Svirskis, 2004

8

Environmental Benefits of Grassland Management

CONTENTS

This chapter looks at the benefits that is being given by the environment to the grassland management. In this chapter it is explained that what are the advantages of the herbicides or what are the advantages of using chemical method for the protection of grassland.

With this it also explains the importance of restoring the fire to native grassland, under which benefits of management of fire and the safety measurements that should be taken for the grassland management are being explained in detail.

As there are several benefits of reseeding, so these benefits basically include drainage, timing factor, spraying, etc. are being explained in this chapter. With this there are several other factors are being explained that are useful for the management of the grassland.

The benefits of grazing animals are being explained which usually contains fire hazards for reduction, plant life benefits and wildlife benefits.

8.1. INTRODUCTION

Grasslands are considered to be one of the major ecosystems of the world while covering around one- third of the terrestrial surface of the Earth (Suttie et al. 2005, Lemaire et al. 2011). Globally, the extensively managed grasslands are recognized for their high biodiversity (Habel et al. 2013), and most of the time they contribute to the agricultural production, along with other rangelands, through livestock grazing on forage which cannot be used by humans directly (Erb et al. 2016).

There are three major types of grasslands ,which can be differentiated within the agricultural production systems and they include natural, semi-natural and improved grasslands (Bullock et al. 2011, Lemaire et al. 2011).

Natural grasslands form the grassland biomes. These are natural areas mainly created by processes which are related to climate, fire as well as wildlife grazing (Parr et al. 2014), and are also used by the livestock.

The semi-natural grasslands are the product of human management. This requires livestock grazing or hay-cutting for the purpose of their maintenance and generally, it will be encroached by shrubs and trees, if taken out of production (Queiroz et al. 2014).

The improved grasslands are pastures and this is the result formed from plowing as well as sowing agricultural varieties or non- native grasses having high production potential. Usually, they are artificially fertilized as well as maintained by using intensive management (Suttie et al. 2005, Pilgrim et al. 2010).

There are two main types of grasslands and they are considered to be important because of the importance that they have for biodiversity, their decrease in areas considering the whole world, and the fact that their full capacity in order to deliver the ecosystem services as the part of agricultural production systems which is not much appreciated.

Natural and semi-natural grasslands are collectively referred to as grasslands.

These have played an important role in the livelihoods of the people for millennia as areas ,which produce fodder for the animals (Emanuelsson. 2009).

The requirement for greater meat and dairy production is not resulting in an increase in the grassland areas which is used for the purpose of grazing but is used for the production of fodder on cropland and improved grasslands.

This is done despite the fact that the growing human population and increased per capita consumption of animal products have caused an increasing demand for animal products (Naylor et al. 2005, Erb et al. 2016).

By the year 2050, it is predicted that there will be increase in the competition for land and other resources for the agriculture (FAO 2009, Smith et al. 2010), this will further be accentuated by ongoing climate change (Harvey and Pilgrim 2011, Röös et al. 2017).

This has resulted in an increased emphasizes on food production as well as food security in the science and policy (for example, Godfray et al. 2010, Foley et al. 2011, SCAR 2011, Poppy et al. 2014). Most of the time, this invokes the concept of sustainable intensification, in terms of the increasing food production on existing cropland (for example, Godfray et al. 2010, Garnett et al. 2013).

During the last century, the grasslands have declined worldwide (Egoh et al. 2016). The main reason for this is the conversion to arable land for the production of animal feed crops and contrastingly, the lack of management as well as abandonment (Queiroz et al. 2014).

Let us consider the example of southern Africa. More than 20 percent of the grassland biome has been cultivated, around 60 percent is irreversibly transformed to some other land uses (Fairbanks et al. 2000). From the remaining, most of the part is used as rangeland for livestock (O'Connor and Bredenkamp, 1997).

In northern Europe, over 90 percent of the semi-natural grasslands have been lost since the 1930s (Eriksson et al. 2002, Bullock et al. 2011, Pe'er et al. 2014). Considering the region of North America, it can be observed that 80 percent of the central grasslands have been converted into cropland (Foley et al. 2005, Suttie et al. 2005).

In a similar manner, the conversion of more than 43 million hectares of the Eurasian steppe has been done into cropland, and around 60 - 80 percent of the grassland area in the region of South America is degraded.

Grasslands have a high conservation value and also, it supports the food production. Grasslands also can be important contributors of ecosystem services (Sala and Paruelo 1997, Pilgrim et al. 2010, Bullock et al. 2011, Lemaire et al. 2011, Lavorel et al. 2013, Werling et al. 2014, Burrascano et al. 2016, Modernel et al. 2016).

However, in the framework of ecosystem services, grasslands have been considered under-appreciated (Frélichová et al. 2014). Sometimes, grasslands have been combined with some other rangelands which include shrubland, deserts and savannas (for example, Sala et al. 2017).

In a similar manner, the grasslands have also received substantially less attention in a number of ecosystem services framework in comparison to other production systems, like the forest (Gamfeldt et al. 2013) and cropland (Robertson et al. 2014).

In the global policy discussions related to the ecosystem services, the grasslands have been largely neglected. Rather it should be highlighted regionally in South Africa (for example) (Reyers et al. 2005, Turpie et al. 2008, Egoh et al. 2016) and the UK (Bullock et al. 2011).

Grasslands have local importance in order to maintain biodiversity and food production. The grasslands also have its effect on the ecological processes at the landscape (such as pollination), regional (such as water regulation, recreation) and global scales (such as climate regulation).

The grassland management is the use as well as the care of grasslands. Along with any agricultural crop, it is important to care for the grasslands in order to keep them productive. The management of plants, animals, equipment and practices which are required for the successful use of grasslands in an agricultural operation is all included in the grassland management.

All the grasslands are said to be natural resources and so, they require good conservation practices. The grassland conservation is referred to as the wise use of grasslands and other natural resources ,which are found on them in order to make sure about the long -term productivity and sustainability. The main aim of good management is to achieve the highest productivity while maintaining the system which provides it.

When the system is pushed being the sustainable limits then it will cause the system to break down. Plants ,as well as soil, will not suddenly disappear but when the grasslands are overgrazed or otherwise abused then it will tend to decrease the productivity over time. When the grasslands are properly managed then they maintain their productivity and even can increase it by helping to build healthy plant communities.

Along with this, when the grasslands are well - managed then they conserve soil, water and wildlife. Research as well as development into improves practices is considered to be a continuing effort and the grassland managers is required to remain informed about the new techniques. Those techniques which can improve their operations. There has been the development of the special grazing systems for livestock which in turn protect the grassland and also, improves the production.

The main aim of the management of grassland is to secure the maximum productivity of livestock without having any detrimental effect to the productivity of grassland.

Therefore, it becomes important to ensure that the grassland under consideration is kept at its peak level of productivity for as much time as much as possible. A decrease of herbage productivity can be because of the bad drainage or lack of sufficient nutrients in the soil or overgrazing.

Recognizing defects is enough easy thing to be done such as bad drainage in soils which is caused by the presence of species which are typical of swampy location like *Cyperus* spp. ,wild rice, *Panicum repens*. In such kind of cases, improvement of the productivity grassland can be done by adopting some measures.

The methods that are to be adopted in order to improve such areas will be mainly mechanical and further designed for removing the excess water from the soil. Nutrients ,as well as physiological defects, can be caused because of the lack of one or more of the important plant nutrients.

When the grass cover is removed continuously and consequent leaching is done then it can deplete the soil of plant nutrients so much that they can become the limiting factors in the growth of vegetation. In such a sort of cases, the restoration of productivity of grassland can be done by using the fertilizers in order to supply efficient elements.

There can be a number of methods which can be used in order to restore the productivity of grasslands or to improve the condition of grasslands and that include:

- Weed control (that is Bush control)
- Enclosure of grasslands
- Burning of grasslands
- Conservation of soil and moisture
- Grazing management
- Reseeding

8.2. ADVANTAGES OF HERBICIDES OR CHEMICAL METHOD OF WEED CONTROL

1. The herbicides can be used as pre - plant and pre - emergence treatment and so, can prove to be helpful in controlling weeds, even before they emerge from the soil. This happens so that crop can germinate as well as grow in an environment which is free of weed or having minimum competition during the tender stage of seedlings. With some other methods of weed control, this is not possible.
2. Herbicides can be useful in broadcast sown and narrow spaced crops. They are very effective in reaching every weed. There are other mechanical methods that are not so effective in such crops.
3. The mechanical methods are effective in wide -spaced crops so as to control the weeds in rows. But it leaves behind the intra - row weeds. Herbicides reach to all places and so, helps in controlling the weeds both inter- row weeds and intra - row weeds.
4. There are weeds which have similar morphological characters like crop and so, are escaped from the mechanical method. But now there are herbicides that are available. The herbicides can kill such weeds without damaging the crop.
5. After the application, for a considerable period, the herbicides withhold the weeds. The weeds tend to grow back soon after applying the mechanical methods.
6. By using the translocated herbicides, deep- rooted, vegetatively propagated weeds can be controlled. There are mechanical methods such as weeding or hoeing. These methods are not so effective so as to control weeds. There are times when the suitable

combination of mechanical methods (deep plouwing or digging) along with the chemical methods prove to be more effective in order to control such weeds.

8.2.1. Limitation of Herbicides (Chemical Method)

1. There must be some technical knowledge while using the herbicides like regarding the choice of particular herbicide, or the time of application, safe dose method of application and so on for any particular crop.
2. If over and under-dose of herbicides is given then that can make a marked difference between the success or failure of controlling the weed.
3. There are certain herbicides which have long residual effects and such herbicides can limit the choice of next crop while performing the crop rotation.
4. The neighboring crops can be harmed because of the herbicides drifts. Let's consider one example. Ester form of 2,4, -D can harm the neighboring crops that can be of cotton, okra, soybean ,etc.
5. There can be environmental pollution because of herbicides.

8.3. RESTORING FIRE TO NATIVE GRASSLANDS

Fire can be used in a safe manner in order to restore the native grasslands. It can also provide a number of benefits to the people as well as nature. Before the period of time when the Europeans settled in the Great Plains and upper Midwest, there were vast prairies and savannas which were once covered about 110 million acres in Minnesota, North Dakota and South Dakota. These grasslands were burned on a regular basis.

Figure 8.1. An image representing the fire that is occurring in the native grassland

Source: Image by Flickr

There were many causes for fires such as lightning strikes, where there were many which were started by Native Americans. They caused such fire for different reasons such as to clear the land for agricultural purposes, for improving the grazing and forage for game species, direct game migration and also, to clear the brush in order to ease travel or prevent hostile forces so that they will not approach unnoticed.

Fire is said to be a natural part of the grassland ecosystem. It helps in maintaining its health as well as vigor. It helps in warming up the soil and further reduces the leaf little which accumulates each year. This ,in turn, allows the sunlight to penetrate. Warming the soil helps in increasing the microbial activity.

This microbial activity releases nutrients from decaying the plant material which new grasses and flowers require for growing.

After a fire, the blackened fields get quickly revived with new, green grasses and also, many showy wildflowers. There are a number of species that thrive with fire and those include big bluestem, purple coneflower and blazing star. Indeed, there are a number of plant species that are found in Minnesota, North Dakota and South Dakota which would be lost without it.

8.3.1. Fire Provides Many Benefits

The fire can be used in a safe as well as an ecologically appropriate manner in order to maintain the grasslands and their plant and animal diversity. There are a number of benefits which can be caused because of fire to people and nature and some of them are mentioned below:

- The fire is sometimes used in order to control the trees, woody shrubs and invasive species (there are times when it is used in combination with cattle or bison grazing). It helps in keeping the grasslands healthy.

After a fire, the grazing animals are attracted to the lush re-growth of grass and therefore, they tend to focus their grazing in that burned area. As new areas are burned, the grazers tend to shift to an area which was recently burned. In this way, it allows previously burned areas to recover.

If the fire is applied in a regular manner then it can reduce the intensity of an unplanned fire. Leaf litter can build up if there will be no fire over time and so, it will provide lots of fuel. If there is an unplanned fire, then it will be much more severe as well as harder so as to control where a lot of fuel has built up.

- Before the arrival of the settler, the native Americans did what is done now. Today, the land managers use fire for maintaining the grasslands for wildlife which include elk and other game species. The healthy grasslands provide food and also, it helps in covering the wildlife all year long. Also, they can provide nutritious forage for cattle.
- Also, the fire can be used in order to increase wildflower diversity in the grasslands. This ,in turn, proves to be beneficial for the butterflies, moths and other pollinators, as it provides a variety of food sources throughout the whole growing season. Different types of vegetation and structure are provided by the diverse grasslands. These are the vegetation and structure which the pollinators need during different times in their life cycles.
- Fire is said to be important in order to maintain the habitat of birds which are dependent on the grassland. These birds include dickcissels, meadowlarks and chestnut-collared longspurs, whose populations are decreasing dramatically across North America and some other parts of the world.

8.3.2. Fire is safe

In order to apply fire, there is a team of highly skilled and experienced professionals and volunteers which is required. There is a 'burn boss' which organizes and also, supervises the event around a written plan and schedule.

There are strict safety procedures that makes sure about the safety of the team members, nearby residents and private property. It is important to check the conditions before a fire is lit which includes weather, wind and drought factors.

8.4. EFFECTS OF ENCLOSURES ON VEGETATION RECOVERY AND SUCCESSION

Vegetation recovery is said to be an economic measure and also, it helps in the development of the local economy. This is done by providing long - term benefits. However, in the decertified areas, the survival of vegetation requires particular conditions, but if in case, such requirements are not met, then it becomes difficult to preserve the vegetation.

The vegetation succession,animals stocks and human interference are the factors that can easily influence the restoration of vegetation. According to the adaptation capabilities of individual plants, there can be variation in the succession, structure and evolution or devolution of plant communities (Whittaker 1972; Watkins and Wilson 1994).

There are conservation measures, which can be helpful in preventing further desertification of grasslands, if they are implemented. These conservation measures can include enclosure practices, Grain for Green policy and control of stock farming on grasslands.

Figure 8.2. An image representing the growth of vegetation after fire.

Source: Image by Earthshots USGS

There have been a number of different studies that have been carried out recently on the effects of duration or method of enclosure on the recovery of grassland ecosystems found in arid lands and deserts (Wang et al. 2001).

The findings of such studies include a remarkable increase that have taken place in the productivity on the lands where the installation of enclosure facilities have been done (Shariff et al. 1994; Biondini et al. 1998; Fuhlendorf et al. 2002; Barger et al. 2004).

There can be positive effects of regeneration of the vegetation on biodiversity (Asefa et al. 2002; Abebe et al. 2006), soil fertility (McIntosh et al. 1997; Su et al. 2005; Mekuria et al. 2007), and furthermore, helps in reduction of soil erosion (Descheemaecker et al. 2006).

There are long - term experiments that can be done in order to investigate the impact of grazing and some other disturbances, during the rehabilitation of desertified grassland, on the longitudinal vegetation and soil changes (Verdoodt et al. 2010).

There are basically six classifications of grasslands namely mountain meadow steppe, typical steppe, meadow steppe, morass, meadow steppe on hills and sandy grassland vegetation. From these types of grasslands, it has been observed that the sandy grassland vegetation is the worst ecological environment. The structure of plant communities changes remarkably over time which in turn are the results of their adaptation to environmental changes (Lv and Lu 2006; Liu et al. 2008). However, the degeneration of the grassland ecosystem takes place further simplifying the composition of plant communities.

The installation of the enclosures is done in order to exclude human activity as well as animal grazing for a specified period of time. Also, it is considered to be a well - known management tool in order to restore the grassland, degraded land and decertified land ecosystems (Wang et al. 2001; Verdoodt et al. 2010).

The enclosure facilities when installed then it boosts the growth of plants. Because of the little human interference, it helps to increase plant communities in a short period of time. this helps in increasing the soil organic matter content, which retains the soil moisture and furthermore, prevents the adverse wind erosion.

Bringing it all together, this helps in fostering the biological diversity and also, it contributes to forestry and stock-farming development.

8.5. ENVIRONMENTAL BENEFITS OF RESEEDING

There is one main advantage of the reseeding grassland and that is that it will increase both the quantity as well as the quality of grass grown. There can be variation in the estimates but new reseeds should grow in the region having 20 - 40 percent more grass. Importantly, a lot of extra growth will be there in the spring and autumn.

The animal performance increases, responsiveness to nitrogen is improved and there will be a great opportunity to tackle the weeds which are all important additional benefits. The importance of a successful reseed is said to be important.

8.5.1. Drainage

It is important to address the issues first before spending a lot of money on reseeding any drainage. Cleaning of the drains should be done and if shores are required, then they are required to be installed.

Ploughing can be a process which in turn helps in drainage and leveling badly damaged fields. But it is important to be aware that the 'good soil' having high fertility is being buried and this process can take a number of years, so as to build them back up again.

8.5.2. Timing

Similar to the rule, in the early growing season, reseeding is done in a much better way. The reason for this is that post- seeding management, which includes grazing off the sward and post spraying, can be somewhat difficult once the later period in the year is considered.

8.5.3. Spraying

If there is a young grass plant and it cannot be very strong and also, it will struggle badly with the competition from weeds. Therefore, it is important to spray off the old sward.

There are many products of glyphoshate and many are effective but the rates require to be checked because they can differ a lot. A number of products may need between 7+ days so that the chemical gets fully absorbed. After the period when the plant is dead, the sward can be cut for silage or grazed.

Fertility

For successful reseeding, it is important to ensure an adequate amount of Phosphorus (P), Potassium (K) and Lime. There can be many ways but the most accurate way is to judge the requirement of the soil. For that purpose, a soil test can be done, but then this will take 2 weeks to get the results, so it is important to plan it in advance. If the plowing is being done, then it is required to wait till the ground is plowed.

If in case, the soil is not being plouwed and a min-till method is been used, that is power harrowing, then lime can be applied in order to counteract the acid which will be produced as the old sward decays.

8.5.4. Fine and Firm Seedbed

There can be times, when a lot of debate can take place about to plow or to use min-till. This decision will depend on many factors which include cost, stoniness of ground, equipment available ,etc. There are a number of methods that can give excellent results but the basic requirement does not change which is fine firm seedbed.

The field should be rolled after sowing, so as to make sure about the good soil to seed contact and preserve the moisture in the soil. If in case, clover is important then it is important to ensure this is on the surface.

8.5.5. Post Spraying

After sowing, chickweed, redshank and docks commonly emerge. It is important that these should be sprayed, preferable with a clover safe spray, 4 - 5 weeks after sowing.

There are pests such as fruit fly and leather jackets and they can attack reseeds. These should be sprayed early, if present. This emerges as a problem when there is a lot of trash left on the surface.

Reseeding is considered to be an expensive investment but this process can deliver excellent returns. However, it is important for the investment to succeed attention to detail and further, giving the young delicate grass plant every chance so that they can establish.

8.6. HOW DIRECT SEEDING BENEFITS THE ENVIRONMENT

The method of direct seeding, or no-till farming, provides producers with an eco-friendly as well as a sustainable alternative to conventional tillage.

As per the advancement of agronomy, there are many people who state to realize just how harmful traditional crop farming techniques can be. No-till farming does improve cropland as well as it also helps in keeping the waterways clean while cutting carbon dioxide production and dust pollution.

8.6.1. Direct Seeding Leads to Better Air Quality

The no-till farming helps in improving air quality in a number of ways. When a direct seeding method is used by the farmers, then it can reduce the use of tractor on an hourly basis per acre by more than 200 percent. This means that the less fuel burned and a few emissions. Also, tilled cropland has a much higher decomposition rate, which in turn releases carbon into the atmosphere.

No-till cropland has the opposite effect, having constant plant cover. It absorbs much more carbon in comparison to what it releases and this helps in offsetting the pollution from cities. One more problem with the conventional tilling is that a number of fields are over-fertilized with nitrates. Over time, the nitrates tend to build up and further break down into nitrogen and nitrous oxide. Both of these substances then easily disperse into the atmosphere.

In a number of tilled fields, dust pollution is said to be an issue. This happens because the soil is bare for most of the year and hence, the dry weather results in dust bowl conditions. No-till fields are more firmly packed and so, always have a cover of vegetation. This prevents the problem of wind erosion.

8.6.2. Direct Seeding Replenishes the Soil

For a number of years, the conventional wisdom taught that the tilling of plant matter for the previous year into the soil for a quick decomposition helped in improving the soil quality. Now, it is known that the reverse is also true. Tilling is done from a long time; wind erosion has removed much of the valuable topsoil which provided a lot of harvests.

The residual crop stubble is left on the surface because of the direct seeding methods, and this protects th soil beneath from wind and water erosion. Year after year, there are new layers of organic material building and further decomposes slowly. Along with this, the direct seeding provides a chance to build rich loam which requires no fertilization in order to keep the crops healthy.

Also, no-till farming helps in building a healthy soil structure. When the soil is healthy, then it has better water retention and also, it contains a wealth of living organisms that can further break down the organic matter into vital nutrients.

8.6.3. Less Pollution Caused by Runoff

The conventional farming methods tend to pollute water sources and this is done by allowing the topsoil to wash into streams, wetlands, lakes and groundwater. Most of the time, it carries residual fertilizers and pesticides along with it. Considering the surface water, it is said to cause algal blooms which ,in turn, damage the fragile ecosystem.

Fertilizers ,as well as pesticides, are said to be responsible for levels of nitrates and phosphates which are dangerous to humans and animals. When direct seeding is implemented, then the year - round plant cover prevents water erosion, whereas, it decreases the use of chemicals which in turn results in fewer chemicals filter into drinking water.

8.6.4. Importance of Soil and Water Conservation

Soil conservation has become an important part of land use. Also, it has support within a social as well as economic environment which is conducive to the maintenance and improvement of soil capital (Dudal, 1981).

The main goal of the soil conservation is to obtain the maximum sustained level of production from the area of land provided while maintaining the soil loss below a threshold level. Theoretically, this allows the natural rate of soil formation in order to keep pace with the rate of soil erosion (Morgan, 1995).

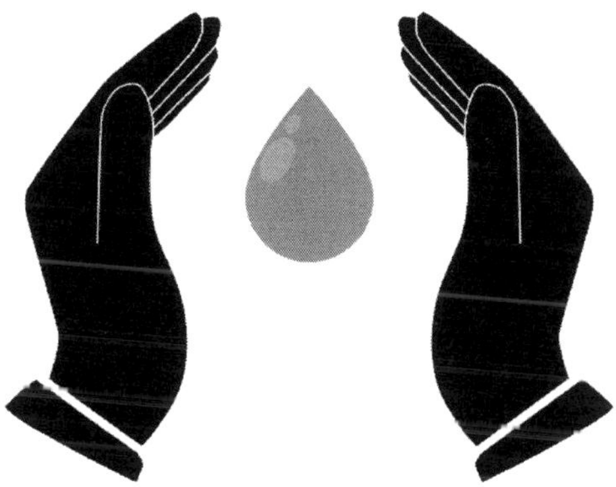

Figure 8.3. An image representing the importance of conservation of water.

Source: Image by Wikimedia Commons

There are primary principles of controlling the soil erosion by water that include reduction of raindrop impacts on the soil, reduction of runoff volume and velocity and increment in soils resistance to erosion (Troeh *et al.*, 1980). There are two technical means in order to achieve these principles of soil as well as water conservation and these include the barrier approach and cover approach (Young, 1997).

The barrier approach is used in order to check the runoff and soil removal by using the means of contour - aligned barriers. These can be terraces like the soil bunds, stone bunds, hillside terraces and grass strips or hedgerows.

The barriers help in diverting the runoff into safe channels, such as grassed waterways and also, reducing it by promoting the infiltration. The introduction of the barriers is done to many new areas of Ethiopia and they also been practiced in a traditional manner for a considerably long period.

In the humid areas, the function of terraces is to decrease the length of hillside slopes and therefore, reduce erosion and furthermore, allows for sediment to settle. Whereas, in drier areas, it serves to retain the runoff as well as increase water available for the growth of plants (Schwab *et al.*, 2002).

There can be a conservation technique which can be regarded as successful if it helps in reduction of the rate of soil loss to less than 20 percent of rate without conservation and to less than 10 t ha^{-1} $year^{-1}$ which is commonly referred to as a 'tolerable' rate of erosion (Young, 1997).

8.6.5. Role of Soil and Water Conservation in Reducing Soil Loss and Retaining Moisture

There are fundamental roles of Soil and Water Conservation structures. It includes to considerably reduce soil loss and also, its consequences.

Considering the practical manner, the loss can be reduced by the structures and this loss is not only soil particles but it is also the most important plant nutrients and applied fertilizers. The measures for Soil and Water Conservation are identified as the first line of defense which acts as a barrier most of the time because of the creation of obstacles against surface runoff.

The main barriers are a channel or basin and embankment of structures. There is a reduction of slope length between structure and it also helps in the reduction of the volume of runoff and therefore, it reduces soil loss. Most of the structures develop in a gradual manner in order to bench and decrease the slope gradient as well as the velocity of runoff.

Figure 8.4. An image representing the importance of soil conservation.

Source: Image by Pixabay

Because of these characteristics of the structures, Tenge *et al.* (in the year 2005) reported that the soil loss is reduced by the grass strips, bench terraces and *fanya juu* by 40, 76 and 88 percent, respectively in comparison to the land without those structures.

As per the report provided by Herweg and Ludi (1999), it can be said that in the Anjeni area of Ethiopia, the soil loss is reduced by 40 percent by the graded soil bund. Whereas the soil loss is reduced by 50 percent by graded *fanya juu* in comparison to the untreated plots.

According to Tesfaye (in the year 2008), the annual soil loss from croplands with level soil bunds is reduced by 51 percent in comparison to the control plot. In Debre Mewi, Ethiopia, the soil loss is reduced by 72.9 and 83.7 percent with stone bund and soil bund respectively in comparison to non-treated land (Teshome *et al.*, 2013).

Considering northern Ethiopia, especially in Tigray, the soil loss is reduced effectively by 68 percent with the stone bund specifically at its early ages. There is a decrease in its effectiveness as the depression on the upslope side of the bunds collects the sediment. Therefore, it requires frequent maintenance in order to sustain effectiveness (Gebrernichael *et al.*, 2005).

There was one more study done in the Tigray region and this study found stone bunds can trap 64 percent of the soil which is moved by water erosion (Vancampenhout *et al.*, 2006). Despite the fact that there was a reduction in the soil loss by 47 percent with soil bunds in the experimental site which is established in the central highlands of Ethiopia in comparison to the non-terraced land.

From the terraced site, the absolute soil loss was still high (that equals to 24 t ha^{-1} $year^{-1}$) (Adimassu *et al.*, 2012) and also, it required specific improvements or support measures in order to reduce absolute soil loss to a tolerable range that is recommended (Young, 1997; Schwab *et al.*, 2002).

The soil loss can be reduced by at least 61 percent by the Soil and Water Conservation in the Tigray region (Girmay *et al.*, 2009) and in the Somali region (Welle *et al.*, 2006), in comparison to the untreated site. The experiment was conducted in Burkina Faso which indicated that the stone rows and grass strips alleviated soil loss by 84 percent and 71 percent, respectively (Zougmore *et al.*, 2009).

The channel ,as well as the embankment of physical Soil and Water Conservation structures, enclose the excess water and furthermore, it enhances the possibility of its infiltration which otherwise takes surface runoff. This role leads to the improvement of soil moisture which is determinant for cropping in areas having medium as well as low rainfall.

In the region of Tanzania, the physical Soil and Water Conservation measures (which include bench terrace, *fanya juu* and grass strap) were effective while conserving moisture (26-36%) in comparison to the land without those structures (Tenge *et al.*, 2005). There are yield and biomass gradient which shows the higher value at the upslope of the stone bund and this gradient is attributed to the moisture - conserving role of those structures (Vancampenhout *et al.*, 2006).

In eastern Ethiopia, the second- order stochastic dominance analysis in the Hunde-Lafto area implied Soil and Water Conservation which in turn mitigated the adverse effects of moisture stress in crop production, especially considering the case of unfavorable rainfall (Bekele, 2005).

8.6.6. Effect of Soil and Water Conservation on Improving Crop Yield

The soil system is considered to be a major determinant of crop yields in comparison to plant genetic potential and weather. And it provides for root growth and that is because of the environment (Olson *et al.*, 1999).

Therefore, agriculture production is increased and sustained and this should aim to sustain higher levels of useful biological productivity as well as to make sure that the system is stable enough so that it can maintain the quality of soil (Dudal, 1981).

The key characteristics of soil affect agricultural yield sustainability and the characteristics include nutrient content, organic matter content, water holding capacity, topsoil depth, soil reaction, salinity and soil biomass (Norton *et al.*, 1999).

There is a relationship between soil erosion and soil productivity and that is complex and also, it involves a number of different factors that depend on each other most of the time (Ludi, 2004).

The soil properties are affected by erosion and therefore, there can be variation in the crop productivity with location and management (Olson *et al.*, 1999). In a general manner, the loss of organic matter ,as well as minerals, contain plant nutrients and this influence crop production (Young, 1989; Lal, 1996; Norton *et al.*, 1999; Olson *et al.*, 1999; Ludi, 2004; Alemayehu *et al.*, 2006).

There can be a direct effect of erosion on the production either through the formation of rills as well as the subsequent washing out of seeds or through an accumulation of eroded materials on the germinated crops (Ludi, 2004). However, total soil loss is considered to be the single most important factor while explaining productivity changes (Ellis-Jones and Tengberg, 2000). There can be a reduction in the total soil loss by physical Soil and Water Conservation measures.

8.7. ENVIRONMENTAL BENEFITS OF IMPROVED GRAZING MANAGEMENT

The well -managed pasture can benefit the environment, and this includes reduced soil erosion; better plant vigor, diversity ,and production; improved air and water quality and improved fish and wildlife habitat. The grazing management can be improved and this can result in more complete vegetative cover as well as improved soil structure.

Then this will allow a higher percentage of the rainfall in order to infiltrate the soil where it can be used for the growth of plant instead of running off where it can lead to soil erosion as well as sedimentation problems.

The increase in the ecological processes includes decomposition of manure in a pasture which is highly managed. Then, during the growing season, the nutrients can be recycled a number of times. With improved grazing management, the entire soil quality can be improved.

With the denser pasture vegetation and improved soil conditions, there can be an improvement in the water quality. There is a study done by the University of Wisconsin which showed that pastures are the best crop in order to reduce the runoff, erosion and phosphorus pollution over any other use of land.

There was a similar study done by USDA - Agricultural Research Service, North Appalachian Experimental Watershed, Coshocton Ohio and this study revealed that both surface and ground water in the pastured watershed was just as good or better in comparison to the water leaving the adjacent pristine wooded watershed. This study provides a comparison in an unimproved typical beef cow - calf pasture.

The pasture soils are a terrific biological filter that is used in order to recover nutrients that pass through the soil. –Throughout the year, the grass roots are active and therefore, can recover nutrients in an efficient manner from pasture soils ,which can leach from other uses of land.

The water quality basically emphasis on shifting the pollution to the non-point source. It was really a surprise for many grazers that from an environmental standpoint, how these pastures are good. The grazers are needed to understand that their systems of farming are when reasonably managed which is meant to be the best from water quality and in mainstream agriculture the standpoint of the soil erosion.

Ohio is considered to contain a hydrological research station. In this station agricultural practices can be studied for their impact on soil erosion, groundwater quality and surface water quality. There was some scientist who worked on the Agricultural Research Services. These services are basically operated by the North Appalachian Experimental Watershed which is present near the Coshocoton. This work was done in Ohio with the Ohio state university (2014).

For collecting the data from the variety of the pasture system Dr. Owens was responsible for this from over 20 years. In comparison to the wooden areas, there was a greater number of infiltrations of the rainfall in the pastures.

To avoid or exclude the livestock from the stream, there was a use of fence and water is being provided from another source. By using these processes, it helps the soil from the pasture which gets reduced from 70.4 tons on the 64 acres which is approximately 1.1 ton per acre to 38.4 tons or approximately 0.6 ton per acre.

This pasture includes the slopes that is from 2 to 35 % on the soils. They are considered as predominately silt loam. Though it is not being measured directly, it is apparent that the soil basically loses. These losses much have very high on the stream banks at the time when the animals use the stream as a water source.

There are some other areas that contain about 25 cow -calf pairs that are being grazed rotationally in 42 acres of land which was basically divided into 4 paddocks. Where one of the paddocks are being served as a winter-feeding area. In this area basically, the hay was fed in the month of November through April that is made from one location on the farm.

In this area specifically the soils were of silt loam. The surface with the texture with 0.37 K value and the value of T is of 4. The slope in this area ranges from 12 to 25 percent which is an average of 20 percent. The soil gets loss during winter area was about 1 ton per acres per year which is about 60 percent of the loss that occurs during winters.

This basically compares the soil losses on a similarly measured watershed which is about 2.9 tons per acre per year. This takes place for the conventionally tilled corn which is planted on the contour. By late- March these winter paddocks are being severely trampled. They are said to appear to be severely damaged.

In the month of May, the padlock is basically recovered and grazed. In order to reduce the soil losses, the rotation of the wintering area basically takes place and this would help to have better distribution of the manure nutrients. In the year 1982, the Ohio Natural Resources Inventory is a result that is been shown as a decline in the pastureland soil erosion rates. The erosion rates on the Land Capability Class VII where pastureland was eroding over 5 tons per acre per year in the year 1987.

The pasture system reduces the time that livestock spends in the confinement. Thus, this pasture system reducing the concentrated manure in the control problems. The manure gets more evenly distributed with the management intensive grazing when compared to the feedlots.

Feedlots are where there is a potential manure odor and they also control concerns as well as unmanaged pastures. Unmanaged pastures are were the animals concentrate on manuring and that is near shade or watering sites.

Manures are generally dispersed about the pasture which is in a properly managed system. This creates a low potential for the nutrient runoff when this is being combined with the low runoff rates.

Now, the third pasture system is containing about 35 cow -calf pairs which are being grazed in the area of 59 acres. It was considered that there were 8.5 acres of summer paddock of orchard grass and about 6.25 acres of land which is tall fescue paddocks that is being grazed in the winter. The hay fed is said to be made from the paddocks. There is approximately 0.1 ton per acre of soil which gets loss from this system which is very low.

These pasture systems are receiving about 200 pounds of nitrogen that is applied in the three equal applications in the month namely April, June and August. This rate of nitrogen is responsible for resulting in the surface runoff with the average nitrate-N level of 3 and 4 ppm for the paddock of summers and winters.

It was considered that during winter paddocks the groundwater nitrate level is comparatively higher. In common, the groundwater nitrate level rises above the 10ppm in both cases. It is considered that the winter paddock receives an additional 300 pounds of nitrogen in hay that was basically bought in as a winter feed.

Legumes were further seeded, which in return gives two results that is fescue and orchard grass. After then, the nitrogen fertilizer is being discontinued, but the fertility level of P and K were still maintained. In this case, the numbers of cows are also being reduced from 35 to 25.

It took several years just for the nitrate- nitrogen level to drop, which normally ranges less than 10 ppm. Whereas, the level of nitrate- nitrogen in the groundwater gets dropped, which normally ranges less than 10 ppm in the winter paddocks and less than 3 ppmin the summer paddocks.

Most of the time, to have higher concentrations of the nitrate that must occur in the runoff are sometimes within the few days following fertilizers application. The best management practices that can be used is to avoid applying fertilizers before the major storm events.

In order to reduce the loss of nitrate- nitrogen and to reduce the potential runoff, it important to maintain good soil health and plant residual after grazing.

Pasture provides an increase in the shelter and food of the wildlife as well as of the grazing animals because it contains permanent and diverse plants cover. It is believed that the properly managed pastures can also provide nesting habitat by delaying mowing and leaving the adequate plant reserve for the rapid growth. Most of the ground -nesting of birds and rabbits prefer the management intensive grazing system.

Research is also showing that grazing animals can also be used to manage the vegetation. This research process takes place on the steam of banks in order to enhance the fish populations. There are many studies that have resulted that the management of intensive grazing systems becomes the recommended conservation practice.

They recommend these conservation practices so as to manage the stream banks on the farms in order to control the overgrowth and also to enhance the fisheries. Minnesota, Canada and Wisconsin are doing these studies.

In this process, the fish number gets gradually increase which is about 2 to 3 times higher. In these areas, the cattle grazed in a management intensive grazing system than the places where cattle were totally excluded from the stream.

There are some recent researches that are being done to tell the impact of the grazing animals that have on the streams, this study says that:

- Preferred crossing points

- Bacteria in streamflow
- Alternative water supplies
- Time in the water and voiding
- Grazing cattle and riparian ecosystems

Cattles are generally accessing watercourses that are used for drinking and crossing. But these cattle are rarely seen to seldom lounged in the water as they don in the ponds. Cattles are believed to have the preferred crossing points. They basically prefer solid footing, as they do not like the soft and muddy creeks bottoms.

When there is a good rock bottom crossing, it allows the cattle to cross at the point and to leave other areas alone. Crossing and accessing points should be designed in such a way that it allows an easy crossing and it should also provide limited access for the drinkers.

When cattle are given chance, then they are observed to drink the spring -fed water when compared to drinking from a stream is about 92 percent. Cattle's are considered to generally do most of their grazing within the 800 feet of a water source.

In many cases, designing and spacing when done in the watering tank, help to save the tanks by keeping away the grazing animals from the stream and also increases the grazing utilization of the pastures. In order to have good animal performance, it important to have clean water.

There can be an increase in the gains by 20 percent if the good clean water is provided from the spring or well. It is considered that rotational grazing supports more diverse bird communities. It also helps in gathering more amphibians and they are taken as good as a grassy buffer which is in regard to the bank stability and in-stream habitat.

Grassland soils are being considered as a great reservoir that is used for organic carbon. There is a rise in the carbon dioxide levels in the atmosphere over the past 100 years. This carbon dioxide level is increased from 290 parts per million to 345 parts in the year 1986. This part is so-called a "Greenhouse Effect".

According to the Council on Environment Qual. And Wild explained that the greenhouse effect is basically warming of the earth's lower atmosphere. This happens due to an increase in the levels of some gases which primarily includes carbon dioxide.

The effect of Greenhouse allows the incoming short-wave radiation from the sun to directly reach the Earth. But the greenhouses are considered to absorb the outgoing long-wave radiations from the Earth's surface back to outer space.

It is considered that the top meter of the soil, contains nearly double the carbon amount than that contained in the vegetation it supports and, in the atmosphere, worldwide. Soil organic compound is considered to contain radiocarbon ages in an extremely stable condition. They are considered with an average 1450 year in the top ten inches of the soil to the ages. These ages increase from 2000 years to 12,100 years with the increase of soil depth to every 90 inches.

The plant community that exists on the soil is said to play a big role in the amount of carbon that is stored in the soil. The total organic carbon is twice as abundant in the prairie soil as in the forest soil. All the soil -forming factors are being considered as equal.

It is considered that the grassland soils are better than the forest soil as the grassland soil also contain more organic carbon with the depth. The forest ecosystem is said to contain the better half of the organic matter that is produced in the above of the ground portion. Whereas, the grass ecosystem is said to contain over 90 percent of the organic matter that exists in the roots.

If any initiative supports grassland agriculture, over a long period of time, it is said to support an effective carbon sink. Grasses and legumes basically use the carbon that is present in the atmosphere for building blocks in the plant tissue. Unutilized and undecomposed plant tissues that primarily contain roots arebeing returned to the soil. In this way, they become part of the soil carbon pool.

This process basically helps in reducing the atmospheric carbon dioxide level. A permanent grassland environment is considered to store more soil carbon than cropland agriculture. It is said that the areas of North Central and Great Plains states are considered to contain the organic matter content of the prairie soils that have been decreased to one half to one- third of the original level.

In a country like England, a forest soil is being formerly cropped and is returned to the permanent sod. The soil carbon content that is present of the upper 10 inches get doubled after 100 years. This increment takes place from 1.4 percent to 3.0 percent. It is considered that 50 percent of this basically increases within the first 20 years.

There are following management practices that are being done for the prescribed grazing management:

- Grazing systems that are based on forage availability and demand. Which basically measure the forage available and determine the animal intake
- Proper rest periods between grazing events (10 to 16 days in the spring 28 to 42 days in the summer)
- Have drinking water available close to the grazing livestock (water distribution is an important part of a grazing system)
- Flash grazing riparian areas (12 to 48 hours)
- Proper stocking rates (know the yield potential for the grazing area and adjust for weather patterns)
- Proper time period (avoid soil compaction)

There is some well -managed grazing system that provides a healthy soil. They also help in cleaning the water, air for the wildlife people and fish. Managed grazing system can be a part of portable farms that are being operated by lowering the cost and also improving the herd health and further reducing the labor.

8.8. BENEFITS OF GRAZING ANIMALS

There are several benefits that can be achieved by grazing animals. These important factors can be used for maintaining the balance and for maintaining the diverse ecosystem.

- Fire Hazard Reduction
- Benefits to Plant Life
- Benefits to Wildlife

8.8.1. Fire Hazard Reduction

There is properly management of grazing livestock that helps in reducing the fire hazards. This is done by controlling the amount and distribution of the grasses and other fuels that are potential. Livestock grazing takes place under license system that is based upon the accepted principles of range management.

The wildlife vegetation programs are being incorporated which happened due to consideration for the forage needs that includes deer and other wildlife, erosion control, and the desire for protecting the sensitive environments. The district ,in this case, is considered to have more than 50 years of experience with this system and is basically found to be beneficial for natural resource management.

8.8.2. Benefits to Plant Life

East bay grasslands and oaks savannas are being generally dominated. This domination takes place by an understory of the non-native annual grasses which commonly contain patches of native grasses and forbs. They also carry varying amounts of coyote brush and poison oak.

Grasslands can accumulate large amounts of dead plant material without having natural disturbances. These dead plant material can also reduce the success of the native grasses and forbs. When a long enough period is being provided without any disturbances, then these grasslands will often convert into the coyote brush, being dominated as shrublands.

The shrubs basically inhibit the germination and growth of the native and non-native grassland plants. this is done by shading out the sun and using the available water and mineral resources in the soil.

Livestock grazing is considered to reduce the competition from the non-native plants. This is done, so that other desirable grasses and wildflowers can regenerate and coexist. There are many plants that are being including several rare species. They require grazing to be maintained at a viable population.

the livestock grazing is often used as a tool in conjunction with the prescribed fire, with the careful applications of the herbicides and mowing in order to promote a healthy and diverse grassland environment.

8.8.3. Benefits to Wildlife

When there is well -managed livestock grazing then that will increase the diversity of the habitats that is available to the wildlife species. There are many species that are being included in several endangered species. These species benefit from the vegetation management that is performed by the livestock.

In many areas ,especially in the East bay, the ground squirrels prefer to colonize livestock grazed grasslands. The ground squirrels' colonies in these grazed areas are being supported by the foraging needs of the predators. These predators are like bobcats and golden eagles. At the same time, they are creating time, creating underground tunnels that care being used by insects, reptiles, amphibians and many other small mammals.

There are also many wildlife species that are being protected which include California tiger salamander, California red-legged frog, san Joaquin kit fox, burrowing owl, and badger that utilizes the ground squirrels burrows. This process takes place throughout the year.

The presence of ponds that are being developed for the livestock for watering support for large number of breeding animals or amphibians. They also feed on the abundant insect life that is found in the surrounding of the grasslands.

In many cases, these man-made stock ponds are providing the replacement habitats for the native amphibians. As the natural water features that are once supported them to have been lost to the development. But the proper utilization of the livestock grazing promotes healthier, diverse wildlife population that is present in the park.

8.9. CONCLUSION

Grassland is basically of two types. These two grasslands are considered to be very important. They are important because they have they commonly have for biodiversity, their decrease in the area considering the whole world. There is also a fact that the full capacity of these grasslands is to deliver the ecosystem services as a part of the agricultural population system. This part is not being appreciated much.

There are some natural and semi-natural grasslands that are collectively referred as grassland. They are considered to play an important role in the livelihoods of the people for millennia. This area is being produced for the fodder of the animal. As there are many benefits of grassland management, fire is one of them. As fire can be used in a safe manner by obeying all the rules that are being used for having control fire when needed. This safer manner helps in restoring the native grasslands. It is also providing a number of benefits to the people as well as to nature.Before the Europeans get settled, there were Great Plains and upper Midwest. As there were vast prairies and savannas which are used to cover about 110 million acres in Minnesota, North Dakota and South Dakota. In these areas, the grassland was burned on a regular basis. The main goal that was behind having soil conservation is to obtain the maximum sustained level of production. This production is basically from the area of land that is provided while maintaining the soil loss which can be below the threshold level. Theoretically, this process allows the natural rate of the soil formation to keep pace with the rate of soil erosion.

There is an increase in the ecological process that basically includes the decomposition of the manure in a pasture. These pastures are being highly managed. During the growing season, it is believed that the nutrients can be recycled a number of times. The entire soil quality can be easily maintained and improved by improved grazing management.

REFERENCES

1. Bengtsson, J., Bullock, J., Egoh, B., Everson, C., Everson, T., O'Connor, T., O'Farrell, P., Smith, H. and Lindborg, R. (2019). Grasslands-more important for ecosystem services than you might think. *Ecosphere*, [online] 10(2), p.e02582. Available at: https://esajournals.onlinelibrary.wiley.com/doi/full/10.1002/ecs2.2582 [Accessed 27 Feb. 2020].
2. Donoghue, M. and Adviser, T. (2018). *2018 - Reseeding - Teagasc | Agriculture and Food Development Authority*. [online] Teagasc.ie. Available at: https://www.teagasc.ie/publications/2018/reseeding.php [Accessed 27 Feb. 2020].
3. Dutch Openers. (2014). *HOW DIRECT SEEDING BENEFITS THE ENVIRONMENT | Dutch Openers*. [online] Available at: https://dutchopeners.com/how-direct-seeding-benefits-the-environment/ [Accessed 27 Feb. 2020].
4. Ebparks.org. (n.d.). *EBRPD - Benefits of Grazing Animals*. [online] Available at: https://www.ebparks.org/about/stewardship/grazing/benefits.htm [Accessed 27 Feb. 2020].
5. Ecoursesonline.iasri.res.in. (2011). *Fodder Production and Grassland Management: Grassland improvement and management techniques*. [online] Available at: http://ecoursesonline.iasri.res.in/mod/page/view.php?id=126625 [Accessed 27 Feb. 2020].
6. F, R., Miller, D. and Nelson, C. (1995). *Introduction to Grassland Management*. 5th ed. [ebook] Columbia. Available at: https://dese.mo.gov/sites/default/files/aged-Grassland-Student-Ref..pdf [Accessed 27 Feb. 2020].
7. Hendershot, R. (2004). *Environmental Benefits of Improved Grazing Management - Pasture*. [online] Livestocktrail.illinois.edu. Available at: http://livestocktrail.illinois.edu/pasturenet/paperDisplay.cfm?ContentID=6618 [Accessed 27 Feb. 2020].
8. Park, K., Qu, Z., Wan, Q., Ding, G. and Wu, B. (2013). Effects of enclosures on vegetation recovery and succession in Hulunbeier steppe, China. *Forest Science and Technology*, [online] 9(1), pp.25-32. Available at: https://www.tandfonline.com/doi/full/10.1080/21580103.2013.774124 [Accessed 27 Feb. 2020].
9. Precisionagricultu.re. (n.d.). *Advantages of Herbicides or Chemical Method of Weed Control*. [online] Available at: https://precisionagricultu.re/advantages-of-herbicides-or-chemical-method-

of-weed-control/ [Accessed 27 Feb. 2020].

10. The Nature Conservancy. (n.d.). *Restoring Fire to Native Grasslands.* [online] Available at: https://www.nature.org/en-us/about-us/where-we-work/united-states/minnesota/stories-in-minnesota/restoring-fire-to-native-grasslands/ [Accessed 27 Feb. 2020].
11. Wolka, K. (2014). Effect of Soil and Water Conservation Measures and Challenges for its Adoption: Ethiopia in Focus. *Journal of Environmental Science and Technology*, [online] 7(4), pp.185-199. Available at: https://scialert.net/fulltext/?doi=jest.2014.185.199 [Accessed 27 Feb. 2020].
12. Abebe et al. 2006
13. Adimassu *et al.*, 2012
14. Alemayehu *et al.*, 2006
15. Asefa et al. 2002
16. Barger et al. 2004
17. Bekele, 2005
18. Biondini et al. 1998
19. Bullock et al. 2011
20. Burrascano et al. 2016
21. CAR 2011,
22. Descheemaecker et al. 2006
23. Dudal, 1981
24. Egoh et al. 2016
25. Ellis-Jones and Tengberg, 2000
26. Emanuelsson 2009
27. Erb et al. 2016
28. Eriksson et al. 2002,
29. Fairbanks et al. 2000
30. FAO 2009
31. Foley et al. 2005
32. Frélichová et al. 2014
33. Fuhlendorf et al. 2002
34. Garnett et al. 2013
35. Gebrernichael *et al.*, 2005

36. Girmay *et al.*, 2009
37. Godfray et al. 2010
38. Habel et al. 2013
39. Harvey and Pilgrim 2011
40. Herweg and Ludi 1999
41. Lal, 1996
42. Lavorel et al. 2013
43. Lemaire et al. 2011
44. Ludi, 2004
45. Lv and Lu 2006; Liu et al. 2008
46. McIntosh et al. 1997
47. Mekuria et al. 2007
48. Modernel et al. 2016
49. Morgan, 1995
50. Naylor et al. 2005
51. Norton *et al.*, 1999
52. O'Connor and Bredenkamp, 1997
53. Olson *et al.*, 1999
54. Parr et al. 2014
55. Pilgrim et al. 2010
56. Poppy et al. 2014
57. Queiroz et al. 2014
58. Röös et al. 2017
59. Sala and Paruelo 1997
60. Schwab *et al.*, 2002
61. Shariff et al. 1994
62. Smith et al. 2010
63. Su et al. 2005
64. Suttie et al. 2005
65. Tenge *et al.*, 2005
66. Tesfaye, 2008
67. Teshome *et al.*, 2013
68. Troeh *et al.*, 1980

69. Vancampenhout *et al*., 2006
70. Wang et al. 2001
71. Watkins and Wilson 1994
72. Welle *et al*., 2006
73. Werling et al. 2014
74. Whittaker 1972
75. Young, 1997
76. Zougmore *et al*., 2009

INDEX

C

D

E

F

G